Deep Excavations in Soil

Deep Excavations in Soil

John Endicott

CRC Press
Taylor & Francis Group
Boca Raton London New York

CRC Press is an imprint of the
Taylor & Francis Group, an **informa** business

First edition published 2020
by CRC Press
2 Park Square, Milton Park, Abingdon, Oxon, OX14 4RN

and by CRC Press
6000 Broken Sound Parkway NW, Suite 300, Boca Raton, FL
33487-2742

First issued in paperback 2022

Publisher's Note
The publisher has gone to great lengths to ensure the quality of this
reprint but points out that some imperfections in the original
copies may be apparent.

Visit the Taylor & Francis Web site at
http://www.taylorandfrancis.com

and the CRC Press Web site at
http://www.crcpress.com

British Library Cataloguing-in-Publication Data
A catalogue record for this book is available from the British Library

ISBN: 978-0-367-50738-1 (pbk)
ISBN: 978-0-367-31360-9 (hbk)
ISBN: 978-0-429-31655-5 (ebk)

DOI: 10.1201/9780429316555

Typeset in Sabon
by Integra Software Services Pvt. Ltd.

To
My family and friends

Contents

Acknowledgements

I wish to thank my wife Pamela, and our children, Emma and James, for their enduring support, which has enabled me to sustain my career.

I also wish to thank the many colleagues and former colleagues who have assisted me in the preparation of this book.

John Cheshire
Johnny Cheuk
Robert Cooke
Suman Dutta
Tina Ho
Alan Lau
Roger Lau
Mable Ma
John Massey
Naseer Mahad
Kumar Pitchumani
Mimi Wong
Ryan Yan
Suki Zhang

Illustrations are courtesy of Kishor Jana.

My career has been formed and guided by the many excellent people who have employed me or been employed by me or whom I have encountered professionally along the way, some of whom are mentioned in the book.

For his patience, I thank Tony Moore of Taylor & Francis Group.

Author biography

John Endicott is well known in Hong Kong and South East Asia as a specialist in engineering deep excavations.

Educated at St Catharine's College Cambridge in Mathematics, he received a first class degree in Engineering in only two years, immediately followed by a PhD in just over three years. He was a College Prize Winner, a Scholar and recently he has become a Fellow Commoner.

His PhD research made use of a large centrifuge to measure deformation within cross-sections of model clay slopes as they deformed when computers for engineering were in their infancy. He and two other students shared all the night shifts for a year on the only computer in the Cambridge University Engineering Department at the time. He ventured into analysis of continuua with plasticity and discontinuua. Within four years he was a pioneer of numerical modelling of soil and structure interaction.

Gaining wide experience in the design of steel box girder bridges and a concrete North Sea Oil Platform and other steel and concrete structures, he qualified in civil engineering, structural engineering and geotechnical engineering. As a Chartered Engineer, he has worked generally in ground engineering including major reclamations, slope stability, foundations, tunnels and underground structures in both soil and in rock.

In 1975 he was transferred to Hong Kong and is still based there with responsibilities in territories stretching from India to New Zealand. In a career of nearly 50 years he has worked on over 100 underground railway stations and many more tunnels. During this time, he developed a business in ground engineering with up to 500 staff in Hong Kong as well as others in nearby countries.

He has published nearly 50 papers on diverse subjects, including hillside and rock mass hydrology, ground treatment, performance of foundations, design and construction of underground structures.

As a Fellow of the Hong Kong Institution of Engineers, he was a Member of Council and helped to establish a Geotechnical Division and then a Geotechnical Discipline, and still found time to be a Member of the Town Planning Board of Hong Kong.

John is one of the few AECOM Fellows in recognition of over 40 years of excellence and dedication.

Abbreviations

AAA	Alert Action Alarm Limits
AC	Accredited Checker
AER	Airport Express Railway
AP	Authorized Person
ASCE	American Society of Civil Engineers
BA	Building Authority
BART	Bay Area Rapid Transit
BCA	Building Control Authority
BCR	Building Construction Regulations
BD	Building Department
BIM	Building Information Management
CDG	Completely Decomposed Granite
CFA	Continuous Flight Auger
CIP	Costed implementation programme
COI	Committee of Inquiry
CPR	Court Procedure Rules
CPT	Cone Penetration Test
Cu	Undrained cohesive strength
ED	Expert Determinator
ELS	Earth lateral support
FoS	Factor of Safety
GBR	Geotechnical Baseline Report
GCO	Geotechnical Control Office
GEO	Geotechnical Engineering Office
GI	Ground Investigation
GIBR	Geotechnical Interpretative and Baseline Report
GIR	Geotechnical Interpretative Report
GIS	Geographic Information System
GIU	Geotechnical Information Unit
GRP	Glass reinforced plastic
HKSAR	Hong Kong Special Administrative Region

Mc	Moisture Content
MDE	Maximum Design Earthquake
MTRCL	Mass Transit Railway Corporation Ltd.
ICE	Independent Checking Engineer
ISL	Island Line
JGP	Jet Grout Pile
LI	Liquidity Index
LL	Liquid Limit
LTA	Land Transport Authority
MIS	Modified Initial System
MRT	Mass Rapid Transit
MTR	Mass Transit Railway
MTRCL	Mass Transit Railway Corporation Limited
NEC	New Engineering Contract
ODE	Operating Design Earthquake
OP	Occupation Permit
OPC	Ordinary Portland Cement
Paul Y	Paul Y Construction Company Limited
PI	Plasticity Index
PL	Plastic Limit
PNAP	Practice notes for authorized persons
QC	Quality Control
QRA	Quantified Risk Assessment
QS	Quantity Surveyor
RC	Registered Contractor
RGE	Registered Geotechnical Engineer
RI	Registered Inspector
RSE	Registered Structural Engineer
RSS	Resident Site Staff
SCL	Shatin to Central Link
SIL	South Island Line
SPT	Standard Penetration Test
TAM	Tube a manchettes
TBM	Tunnel Boring Machine
U.K.	United Kingdom
U.S.A.	United States of America

Symbols

Ka	Coefficient of active pressure
Ko	Coefficient of at-rest pressure
Kp	Coefficient of passive pressure
Kh	Coefficient of horizontal sub-grade reaction

c_u Undrained cohesion
p' Isotropic stress
q Deviator stress
K
Σ
$\sigma\varphi\tau\gamma$

Chapter I

What are deep excavations?

To an average person, deep excavation might conjure up thoughts of massive deep open cast gold mines. These can take dozens of years to excavate with more than a million tonnes of earth removed in one day. The largest of these goes down 700m [1], deep enough to accommodate a 230-floor tower without appearing above the original ground level, nearly as big as Burj Khalifa (829.8m) and more than Tokyo Skytree (634m) and Shanghai Tower (632m). Such deep excavations for mining rock ore are generally in undeveloped terrain. Engineering for deep open cast mines requires substantial input from rock mechanics engineers. Deep excavations in soil are generally not as deep as open cast mines and necessitate much more gentle slopes or retaining walls to hold up the sides.

Deep excavations in soil are generally carried out to enable the building of underground structures. One definition of deep excavations is based on the premise that if an excavation were to collapse, the consequences would be serious, such as causing a fatality or fatalities if people were within or adjacent to the excavation. Building Construction Regulations [2] in Hong Kong require a qualified engineer to design lateral ground support for excavations deeper than 1.2m. Soil is heavy. One cubic metre of soil that could collapse from the side of a pit 1.2m deep weighs 1.5 to 2.0 tonnes; it would seriously damage the legs of anybody standing there. Deeper excavations are potentially of even more serious concern. Excavations for six levels of a basement can exceed 18m in depth; the collapse of such an excavation would drastically undermine the surrounding area. For this reason, deep excavations in soil generally require lateral support to the surrounding ground to prevent caving-in and to limit ground movement to acceptably small amounts.

Deep excavations in soil to enable the construction of underground structures are complicated due to the variable types of ground and groundwater that they may encounter. In urban areas the surrounding land must be fully supported, and nearby structures have to be protected, sometimes by limiting ground movement and adjacent building movement to within a few millimetres. Occasionally, ancient monuments or trees with preservation orders must be protected. The sides of a deep excavation need robust Earth Lateral Support, often comprising reinforced concrete retaining walls braced by steel or concrete struts that sometimes, depending on the loads, can be quite heavy. A range of professional input is required at all stages of a deep excavation project. Professional input is required for developing concepts, for planning, for geotechnical studies, for structural design, for programming, for management, for site supervision, for certification of payment and for the resolution of any claims arising from the excavation and construction.

Successful completion of a deep excavation requires several different skills, such as planning, design, permitting, contracts, supervision, completion, payment and resolution of disputes and, if it is to be done well, a capacity for lateral thinking. Probably the man in the street does not realise just how much work is involved in excavating tens of thousands or hundreds of thousands of cubic metres of soil. This book aims to cover the whole process for people to flip through to get a taste for the subject and then to re-read to appreciate some of the world's engineering marvels below the ground. For professional ground engineers, there is a trend towards more and more specialisation. Professional training schemes require chartered engineers to have knowledge about the whole process of civil engineering projects and not to focus on only one part, such as the cutting and pasting of soil parameters into a computer program. Whatever the specialisation, engineers who work on deep excavations and engineers for other associated civil engineering works need to have an appreciation of all of the other tasks that are required for the project and be able to exhibit the independent judgement necessary to know when and who to ask for assistance when required.

Often, construction sites are in urban areas. For reasons of safety, the sites are generally surrounded by a protective hoarding. Projecting above hoarding, intriguing 30m tall construction equipment can often be seen. Gateways in hoarding provide access to sites, and traffic controllers are stationed to admit heavy trucks with equipment or materials and trucks laden with excavated soil to leave the site dripping water after being hosed down to clean the wheels. A glimpse through a site entrance, or through a window in hoarding, can reveal excavators at work, massive shoring being erected between deep walls, various stages of partial construction of the permanent structure and in some cases substantial shoring to protect adjacent ground and buildings from movement.

Deep excavations in urban areas are adopted for multi-level basements for buildings and for underground road, underground rail and drainage infrastructure. Extensive works are particularly required for underground railways. A common layout for an underground railway station is to provide several pedestrian subway entrances down to a public area concourse level with ticketing and access through to paid areas and via escalators and staircases to platforms and rail tracks below. Excavation for a new two-floor underground station is about 15m to 18m deep to allow for a 1m- to 2m-thick base slab beneath the tracks, headroom of say 5m to 7m above the tracks for the railway system, 1.5m overhead ducting, a normal public area headroom of about 3m at the concourse level, a substantial thickness, say 1m, of roof structure and finally 2m or more space for buried pipelines and cable utilities before reaching finished ground level. Stations with cross-platform interchanges require two levels for platforms and are deeper by 5m to 7m. Where several railway lines cross each other, excavations go even deeper. For example, at Admiralty Station, Hong Kong, when constructing the South Island Line and extension works to the existing station, the excavations reached 45m below ground and included excavation beneath existing railway tunnels operating at shallower depths [3].

Railway tunnels between stations are sometimes deeper than the stations. Energy-saving tracks have downhill slopes assisting trains to accelerate on departing from stations and with uphill slopes to assist approaching trains to slow down and stop at the stations. These slopes located to either end of stations were called energy humps. They are no longer so common because the electrical power system for trains can incorporate regenerative breaking, as used on hybrid or electric cars whereby electrical braking puts energy back into the grid system. Between underground stations, shafts are sometimes required such as for ventilation, for drainage or for means of emergency escape. Whereas stations are typically 15m to 25m deep, some tunnels between stations can be much deeper.

With urban growth, more and more deep excavations are being built in urban areas. Many cities, especially in China, India and South East Asia, are getting much bigger, land above ground is getting more and more congested and land for construction is getting very rare. More utilities are needed for the burgeoning populations. In the past, utilities such as cables were slung between poles above ground in an unsightly manner and drainage was often in open channels. Excavating deeply below ground for new utilities is often necessary because of the congestion at the surface and going below ground with unsightly infrastructure is therefore sensible. With high costs for land in urban areas, basements are becoming more and more cost-effective and are also being constructed to greater depth to build more levels of floors. For example, in Jurong New Town, Singapore, new commercial basements extend to six floors below ground level. These include car parking and supermarkets where one can take a trolley loaded with purchases from a shop to the car at 15m or more below ground in air-conditioned comfort. Subways connect from one building to another obviating the need to cross busy roads in the hot open air. Direct connections from basements to underground railway stations have become popular in Hong Kong since 1978, when it was realised that entrances to underground railway stations would have large pedestrian traffic; basements that were connected to underground railway stations became prime sites for retail shops.

In many cities, underground development offers protection from extreme weather with cooling in tropical locations and warmth in cold places such as Helsinki, Montreal and other cities that are frozen in winter. Expansion of cities and buoyant economies means that the citizens expect an improved way of life and an improved living environment. With land at the surface already intensively developed and therefore in short supply, going below ground is a means of developing infrastructure that improves overall city life. Construction below ground can provide infrastructure that otherwise would take up space at the surface, in some cases releasing space that was previously occupied. In congested cities, open space is a much-valued commodity. Underground railways, which were started more than 150 years ago, reduce pedestrian traffic. If Hong Kong, Singapore and other cities had not developed underground railways since the 1970s, the streets would have become more than stagnant with twice as many vehicles. Bangkok has opened some underground lines, and where there used to be unpleasant areas with almost stationary traffic and smoke from two-stroke "Tuk-tuk" engines, there are now attractive city districts, and it is pleasant to walk by day or by night.

Mobility of pedestrians is key to life in a city. Many cities have developed, or are developing, extensive underground pedestrian subways, concourses and malls. These provide needed space for people to do activities and provide protection from the weather and, in the case of Singapore, civil defence. In busy streets, cycling can be a nightmare. Tunnels for cyclists are planned beneath Cambridge U.K., and cyclists in Paris have found routes through basements and subways. In addition to car parking and shopping, there are leisure activities below ground such as cinemas, ice rinks, swimming pools and bowling alleys. In fact, a whole host of activities can go below ground if the demand is there.

References

[1] Twin Creeks gold mine in Nevada, United States. https://en.wikipedia.org/wiki/Open-pit_mining#/media/File:Twincreeksblast.jpg

[2] *Building Construction Regulations Cap 135*, HKSAR Government. www.elegislation.gov.hk/hk/cap123B?xpid=ID_1438402645257_002

[3] Bezzano, M., Smith, S., Yiu, J. & Wiltshire, M. Case Study: Design and Construction Challenges for Admiralty Station Expansion. *Proceedings of the HKIE Geotechnical Division Annual Seminar*, 2017. www.hkieged.org/download/as/as2017.pdf

Chapter 2

How deep excavations are created

Deep excavations are expensive works of civil engineering. They usually come about under a civil engineering contract let to a construction company after planning and design by engineers. The parties that are involved are a developer, usually referred to in contracts as "The Employer," a designer who is usually referred to as "The Engineer," who might also be entrusted to manage the contract, and a construction company who is referred to as "The Contractor." For clarity, I will refer to "an Engineer," using the upper case, when describing the role of an engineer under a contract, and I will refer to "engineers," using the lower case, generally. The contractual role of an engineer and his team for a deep excavation is comprehensive from the start, when planning the excavation, until the satisfactory completion of the works and completion of the contract. In summary, the role of an engineer is to arrange for the site investigation, design and construction to be carried out, to write contract documents for the work and to arrange tendering for a construction contract to be awarded. During construction, an engineer ensures that the construction is carried out properly and that the completed excavation is handed over to the developer of another engineer or a contractor for subsequent work on the site. An engineer is usually responsible to ensure that any claims arising from the works are resolved and to certify payments.

Aspiring young engineers are often attracted to one specific the stage of the process. One engineer might want to be a designer; another might want to be a builder or a manager. Nowadays, deep excavations are generally large enough and of such complexity that there is room and a need for a team of engineers with the necessary range of qualifications and experience. However, to be the Engineer, the person and team leader who is ultimately responsible for a deep excavation, one should know about every task that is required. A leader does not have to be a specialist in all aspects himself. He (to be read as singular person throughout) needs to make sure that all the aspects are dealt with to the required standards and they need to be able to recognise when specialist advice is needed. As a member of a team, in order to fulfil one's role, one needs to understand the respective roles of the other members of the team and how the whole process fits together.

2.1 Types of earth lateral support

Fundamentally, for deep excavations, the surrounding ground should remain stable. Movements of the sides of the excavation must be carefully controlled otherwise nearby property could move or be damaged. Generally, the adjoining ground needs to be fully supported with an Earth Lateral Support (ELS) system during the excavation process. An ELS system generally comprises a robust wall around the perimeter of the excavation and shoring to prevent the wall from moving more than a small amount. Where space is available, and the ground is strong enough, sometimes side slopes can be formed and an ELS system might not be required. For deep excavations in soil that is not self-supporting and where space is limited, an ELS system is definitely required. Especially below the ground water level and in weak soils, the ground has to be supported. Below the water table, the soils may wash into the excavation and weak soils could slump as a mass. Buildings Construction Regulations in Hong Kong[1] require that excavations greater than 2.5m deep shall be provided with adequate ELS. Failure of a 2.5m deep cutting would be potentially fatal or of serious consequence for people working in the excavation. Excavations that are considerably deeper than 2.5m require ELS all the way down. As soon as excavation reaches 2.5m depth, or even before, ELS shall be provided. ELS can be provided sequentially as the ground is excavated, by installing timbering or sprayed concrete to prevent the exposed face of the soil from ravelling, and overall support to the ground can be provided by installing props, in the case of narrow trenched excavations or by soil nails or tie-back anchors in the surrounding ground. However, these methods are not suitable for very weak soils or below-the-ground water tables where the base of the excavation could be unstable. Therefore, generally an ELS system is provided by installing robust walls from the surface before any significant excavating has taken place, and then bracing is installed as the excavation advances stage by stage.

Several techniques have been developed to install earth retaining walls from the surface before excavating the bulk of the soil from the site. The five common methods are driven steel sheet piling, king posts and lagging, contiguous bored pile walls, secant piled walls and diaphragm walls. In some countries where labour is inexpensive, piles can be excavated by hand and filled with concrete, with or without steel reinforcement, and are called hand-dug caissons. Rows of hand-dug caissons have also been used as earth retaining walls for deep excavations.

2.1.1 Driven steel sheet piling

Steel sheet piling has been used for many years as earth retaining structures. Sheet piles are driven into the ground to a required depth in advance of excavating. In the past, sheet piles were driven with concrete or steel drop hammers weighing a few tonnes and were lifted and dropped from a tripod frame or by a crane. The method is very noisy, it creates considerable ground vibration and now it is seldom used in developed areas. Diesel hammers are more efficient than drop hammers and diesel hydraulic hammers are even more efficient, but both types are noisy. Hydraulic sheet pile drivers are much quieter. They are clamped to a row of four or five sheet piles at a time, and hydraulic jacks push down one pile at a time obtaining the necessary reaction from the other piles in the group. The long edges of steel sheet piles have clutches so that adjacent piles are linked together, and if the clutches are in good condition seepage of ground water through them is minimal. Steel sheet piling is manufactured to various sizes and the structural capacity can be increased by welding to them other steel sections such that the two members act compositely. Steel sheet piling is often used as temporary earth lateral support for excavations as deep as about 10m to 15m; in some countries, steel sheet piling can be used as permanent walls for basements. Practical limitations include both the structural capacity that needs to be greater for deeper excavations and the supplied lengths of 15m, meaning that longer lengths require site welding. Steel sheet piles are provided with clutches along each edge that join adjacent piles together; a wall of interlocked sheet piles can provide an impermeable barrier to water. However, when driven into hard ground there is a risk that the clutches could leak or separate and might require sealing by grouting the soil outside the gaps. Steel sheet piles can be driven to penetrate quite strong ground, but driven steel piles can be obstructed by hard objects such as boulders. A common specification for placing rock fill where piles are intended to be driven is to limit the maximum size of the rock fill to 200mm.

2.1.2 King posts and lagging

This type of ELS involves boring vertical holes in a line in order to place king posts vertically at intervals along the alignment of the earth retaining walls. The king posts are usually either steel beams placed in the bored holes or reinforced concrete piles that are cast in the bored holes. As a rare alternative, precast concrete piles are placed in the bored holes. King posts provide the main structural elements of support to the ground. With the king posts in place, excavation progresses, and lagging is placed between the king posts. Lagging provides support to the exposed ground between the king posts. Planks of timber, steel or pre-cast concrete are used as lagging, and they generally span horizontally between king posts and are restrained by the king posts. It is less common to provide lagging by cast-in-situ reinforced concrete. A few projects in Hong Kong have used unreinforced concrete arches, curved in plan spanning between king posts. In firm ground, it is sometimes possible for the ground to support itself by arching horizontally between the king posts; sprayed concrete, also called shotcrete, may be used to support the exposed surface of the ground by arching between king posts and to prevent the ground from ravelling.

Earth support with lagging is pervious to groundwater and cannot be used below the water table in permeable ground. The use of lagging relies on the ground being temporarily self-supporting when it is exposed after excavating and until the lagging is placed and secured. Therefore, this method generally cannot be used in very weak ground. Moreover, lagging provides no support to the ground below the base of the excavation. The individual king posts provide some support to the ground beneath the base of the excavation, but in very weak ground there is a risk of base failure of the ground between the king posts heaving and causing the base of the excavation to rise.

King posts and lagging are primarily used for temporary ELS. An exception is robust walls comprising steel king posts and unreinforced concrete arched lagging being used for the longitudinal walls of the Diamond Hill underground railway station in Hong Kong[2]. When first designed, it was intended that the station would be a cross-platform interchange station. In order to facilitate widening, the longitudinal walls were required to be removable. They were constructed with steel king posts and unreinforced concrete lagging arching between the posts. Forty years later, the station layout was changed to a grade-separated interchange with the new line passing beneath the original line. Widening was not needed. The longitudinal walls were exposed, examined and checked by current standards and were found to be suitable for permanent use.

2.1.3 Contiguous bored pile walls

Contiguous piled walls are formed by constructing rows of bored piles along the alignment for each side of the proposed excavation. The piles are usually concrete piles that are cast in place. The piles may all be reinforced concrete, or the piles may be alternately reinforced concrete and unreinforced concrete. The piles are adjacent and touching each other. The joints between the piles are not impervious and often must be rendered impervious by grouting. A contiguous bored pile wall can be used for both temporary walls and permanent walls. The surface is not smooth and flat, and a secondary internal wall can be used for the sake of achieving a smooth flat surface or to provide a vapour barrier and a gap for drainage if the contiguous bored pile wall is not completely water tight.

The contiguous piles that are used for a wall can be conventional bored piles, continuous flight auger (CFA) piles, soil-mixing piles or jet grout piles (JGPs). Conventional piles are constructed by boring a vertical hole in the ground and filling the hole with plain or reinforced concrete. There is a range of methods of creating bored piles depending on the type of ground and the available equipment. Boring can be carried out by a boring machine, which rotates a cutter into the ground and, when retracted, removes excavated spoil. Boring can also be carried out by sinking a heavy steel casing, with a cutting edge at the bottom, into the ground. The casing is sunk by rotating it with a heavy plant or a vibrator. As the heavy casing is sunk into the ground, the soil within the casing is removed using a clamshell grab or an auger. Boring with a heavy casing in this way is a relatively slow process. Continuous boring in soil is usually faster and can be achieved by lowering a reverse circulation machine down the hole and flushing the cuttings to the surface suspended in slurry of bentonite and water. Bentonite slurry is used for bored piles and is commonly used when excavating for diaphragm walls (as described later). In firm ground, bored holes can be stable and need no support, in which case a bore can remain open while excavating. When boring below the ground water table, a bored hole may be filled with water to balance the ground water and prevent it from flowing into the hole. In weak ground, bentonite slurry may be used to balance the ground water pressure and to support the ground. In very weak ground, a permanent steel liner may be advanced as the hole is bored. The steel liner provides temporary support to the ground while boring the hole, and it provides a permanent smooth surface. Bored piles are formed by placing a fabricated cage of steel reinforcement, if required, in the bored hole and then filling the hole with concrete.

CFA piles are bored with special equipment comprising a long continuous flight auger. The auger is rotated and pressed into the ground and spoil is removed via the continuous flight to the surface. When the desired depth is reached, cement mortar is injected under pressure via the hollow shaft of the auger to the base; as the mortar is injected, the auger is removed. During the construction of each pile, the bored hole remains full of material or excavated soil, or it is replaced by mortar as the filling progresses, thus preventing the collapse of the bored hole. If required, steel reinforcement can be lowered into the mortar-filled hole and may have to be vibrated to penetrate the full depth of the mortar-filled bored hole.

Soil mixing is a method whereby soil is mixed in situ with cement paste and is allowed to set as a weak concrete. The equipment that is used comprises a horizontal blade at the bottom of a hollow steel rod. The rod and blade are pressed and rotated to penetrate the ground to the required depth. The equipment is then rotated and slowly withdrawn while cement paste is injected through the hollow shaft and discharged at the bottom where the rotating blade mixes the wet cement with the soil. The resulting pile is approximately the same diameter as the cutting blade and the pile is not reinforced. The equipment is designed for use in weak soil and is not suitable for very strong soil.

Jet grouting is another method of forming unreinforced piles in weak ground. The equipment that is used comprises a rod with a cutting tool at the bottom. Water is fed at very high pressure through the cutting tool emerging as a high-speed horizontal jet at the tip as the cutting tool is rotated. The jet of water cuts the ground making a suspension of cuttings in water, and the pressure lifts the suspension of cuttings to the surface. The diameter of the cut depends on the strength of the ground and the applied pressure of the water. In weak ground, the diameter is generally of the order of three quarters of a metre making posts with diameters of about one and a half metres. In firm ground, the diameter is smaller. When the tool reaches the required depth, cement slurry is pumped through the tool instead of water and at a lower pressure thereby filling the excavated space with cement, which develops a strength of about 1MPa to 5MPa. In sandy soil, the method can be adjusted to mix the cement with natural sand, and the resulting sand mortar can be stronger than 5MPa.

In order to form a wall, contiguous piles are placed closely in a line abutting each other. The piles are constructed as closely as possible, but gaps between them could leak ground water. A contiguous piled wall can be made impervious to ground water if the spaces between the piles are sealed with grout or if the ground immediately outside the joints between piles is sealed with grout. Grouting between contiguous piles can be done by drilling a hole for grouting down each joint or just behind it. Grout is pumped down the grout hole and, depending on the type of ground and the type of grout, the grout penetrates the ground to a limited extent. When sealing vertical joints between piles, grouting is required to be effective up and down the joint. Sometimes it is sufficient to grout slowly through a pipe that is first lowered to the bottom of the grout hole and slowly raised with the objective of allowing the grout to penetrate the ground all the way up the hole. Often the ground is variable and different types of grout are required to be injected at different rates at various depths. For this purpose, grout holes are lined with a perforated pipe and a double packer is used to inject grout into the ground through the perforations at a series of levels going up or down each grout hole. When an elastic sleeve is located outside the pipe at the levels of the perforations, the system, which was devised by French engineers, is referred to a "tube a manchettes" (TAM). TAM grouting facilitates re-grouting the same hole at the depth of each sleeve. Grout is injected at selected sleeves and the sleeves close after grouting such that the tube can be flushed. After flushing, the grout pipe is clean and unobstructed by grout, and it can be used for grouting again.

Several types of grout can be used, depending on the type of ground to be grouted. Probably the cheapest grouting material is cement. As long as the amount of water used is not excessive, cement sets as a strong impermeable material. However, cement particles have a diameter of about 0.1mm and will not permeate soils with pores between soil particles smaller than about 0.5mm, which includes almost any soil with more than about 35% fines (fines are silt and/or clay). Cement grout only permeates clean coarse sand, gravel, cobbles and boulders. Cement grout can be forced into the ground under sufficiently high pressure by rupturing the ground and forming veins, lenses or "bubbles" of cement in the ground. The lenses or bubbles are called "clacquages." Clacquages are formed of solid impermeable grout while the remainder of the ground remains not grouted. For permeable soils with small pores that need to be made less permeable, chemical grouts, especially water-soluble grouts, are used. A solution of sodium silicate in water has viscosity the same as water and can penetrate wherever water can penetrate. It sets as a soft but waterproof gel; it is commonly used for reducing the permeability of fine-grained soils and for sealing gaps between contiguous piles.

2.1.4 Secant piled walls

Secant piles are also used for constructing walls by forming a row of bored piles in the same way contiguous piles are built in a row. Secant piles are also bored concrete piles that are cast in place. Unlike contiguous piles that are constructed touching each other, secant piles overlap in order to create a wide contact zone between adjacent piles. Secant piles are alternately primary unreinforced weak piles and secondary reinforced concrete piles. The primary piles are formed from plain concrete, cement-stabilised soil or jet grout. The secondary piles are conventional bored piles usually filled by strong concrete with steel reinforcement. The secondary piles are located between the primary piles. When boring the secondary piles, the secondary piles are cut into the first piles. The secondary piles are the main structural elements, and the primary piles are intended to support the ground between the secondary piles and to provide a water seal between the secondary piles. The layout of secant piles is illustrated in Figure 2.1.

Usually, the primary piles are cast with low-strength concrete because the primary piles are to be cut by the secondary piles. When primary piles are too strong, the drilling for the secondary piles tends to go off-line

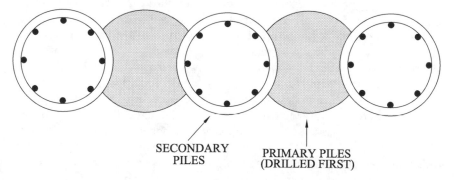

SECONDARY PILES   PRIMARY PILES (DRILLED FIRST)

Figure 2.1 Secant bored piles.

resulting in an unacceptable lack of verticality and gaps between the primary piles and the secondary piles, and the pile-boring equipment can become stuck. Secant pile walls are used mainly for temporary walls because of the low strength and lack of durability of the primary piles.

2.1.5 Diaphragm walls

Diaphragm walls are so-named because the panels of reinforced concrete usually have steel reinforcement with structural capacity both horizontally and vertically, and each panel can act as a structural diaphragm. This type of wall is the most robust currently and widely used for deep excavations in soil. Because of their large structural capacity, diaphragm walls are used for very deep excavations, for example in shafts as deep as 60m. Because of their robust construction, diaphragm walls are used for many temporary walls and for permanent walls to underground structures. Diaphragm walls are generally made from cast-in-situ reinforced concrete, although precast panels have been used and have the benefit of a smooth formed inside face for permanent use, and some cast-in-situ walls have been strengthened by post tensioning, although this is not a common practice. Diaphragm walls can be built within a wide range of capacity from plain concrete walls about 600mm thick up to reinforced concrete walls more than 1500mm thick. Larger capacity walls can be made from joined panels, such a Tee-panel made of two panels joined at right angle. The edge of one panel is joined to the middle of the face of the other panel. Steel reinforcement for the two panels is linked together and the concrete for the two panels is cast monolithically, in one go, thereby forming the shape of the letter "T" in plan. Diaphragm walls can be used, say for shafts, in a circular layout requiring no bracing because of the circular arching action of the completed wall. For large diameter shafts and to allow for out-of-alignment wall panels, sometimes waling beams are added internally to brace wall panels. For longer excavations, the layout of diaphragm wall panels can follow a series of curves as if they are overlapping circles in plan with transverse beams at the cusps. Such layouts are called, descriptively, caterpillars.

The method of construction for diaphragm walls commences before general excavation. Shallow trenches, of the order of 1.2m deep, are constructed with reinforced concrete guide walls to either side, along the alignment for the diaphragm walls. These guide walls serve to provide the location for excavating deep trenches and provide support at the edges of trenches where heavy plant will be operated. Trenches for the wall panels are excavated in short lengths. The narrowest panel is equal to the width of the diaphragm wall-excavating equipment, which is typically 2.8m. Individual panels are often excavated by two or three passes of the excavating equipment and can be up to about 6m long. The trenches for panels in soil generally extend at least to the final depth of the excavation and generally more to be embedded below the final excavation level. While being excavated, trenches are kept filled with slurry. When a trench has been excavated to the required depth, steel reinforcement is lowered into place and the wall panel is filled with concrete by the tremie method. The method utilises a pipe that is lowered to the bottom of a slurry-filled trench. Concrete is poured inside the pipe to reach the bottom of the trench without mixing with the slurry. As concrete is poured, the pipe is progressively lifted keeping the bottom of the pipe just below the top of the concrete. As the height of the concrete increases, slurry is displaced and removed from the trench.

Diaphragm wall panels are excavated either by grabs or by cutters. Grabs are made to the size of the thickness of the wall panels, mainly in the range of 600mm to 1.5m thick, and when the grabs are opened, they are typically 2.8m wide. Usually grabs are operated suspended from heavy crawler cranes, typically 90 tonnes to 100 tonnes capacity. Sometimes grabs are operated suspended from special gantries, such as when they are required to operate under limited headroom. Some grabs are operated from Kelly Bars (i.e. heavy telescopic steel tubes suspended from a crane). The depth of excavation is limited by the extension of the Kelly Bar. Grabs that are suspended from cables can extend to greater depths. The jaws of grabs can be operated by cables or by hydraulics. The closure of the jaws is stronger when using hydraulics, and weak rock can be ripped by hydraulically operated grabs. Grabs are used to excavate soil and can excavate weak rocks that can be ripped, but they are obstructed by rock or other strong buried materials such as concrete. When such strong obstructions are encountered, heavy chisels may be used to break up the obstruction and the fragments are then removed by using the grabs. This tends to be a time-consuming process.

Rotating, hydraulically operated cutting machines are also used to excavate panels for diaphragm walls. Such cutting machines are called "trenchers" or "hydrofraises" and are more powerful and more versatile than grabs. The equipment includes hydraulically driven rotating cutters at the base that can be fitted with teeth that can break up soil and can be strong enough to cut many types of rock and other obstructions. The equipment includes slurry pumps that are used to pump slurry up to the surface with suspended fragments of the broken-up ground. Cutters are operated continuously, and rates of excavation are generally faster than with grabs. Grabs operate by removing the soil as the grabs are lifted to the surface, which is time-consuming. Cutters can work continuously without pauses to lift the spoil to the surface. In order operate the cutters efficiently, trenches filled with slurry should be several metres deep; it is common practice for grabs to be used for a first stage of excavation in the trenches and for cutting machines then to take over.

For both types of equipment, grabs and cutting machines, trenches are kept filled with slurry. Slurry is a suspension of bentonite in water or a solution of polymer. The primary purpose for slurry is to support the sides of the trenches and to limit ground movement during excavating trenches and during concreting. Large volumes of slurry are used, which requires space for mixers, storage tanks, slurry treatment plant and testing equipment. Maintaining the quality of slurry is very important. When using cutting machines, a secondary, but important, purpose for the slurry is to suspend the excavated fragments of soil as it is pumped out from the bottom of the trench.

Bentonite is a thixotropic material. Polymers are manufactured for the same purpose. Thixotropy, creating a gel, is important for stabilising the sides of the excavated trench for diaphragm walls. Natural bentonite is mostly clay called montmorillonite and has thixotropic properties. When it is stirred, it is fluid, and when it is at rest, it sets as a gel. When it is mixed with other soils, the thixotropic properties can be impaired or lost. Bentonite slurry was originally used as a mud for deep drilling in the oil industry for holding particles of rock in suspension for removal from deep drill holes. This ability to hold particles in suspension helps to lift the excavated spoil to the surface, but the spoil has to be removed before the slurry can be returned to a trench for re-use. Removal of spoil requires sieving and repeated cleansing. The main advantage of slurry in trenches is the thixotropic properties that can seal the exposed face of the ground. By keeping the level of slurry above the elevation of the ground water, a positive pressure is applied to the sides of a trench, and the sides are stabilised with a thin layer of gel. If a seal is not formed on the sides of a trench, loose soil can fall into the trench from the sides creating overbreak. Subsequently concrete fills any overbreak and encroaches into the main excavation where it must be cut off later. If the ground alongside a trench is porous, then bentonite slurry can permeate the ground by a short distance and form a gel within the pores, thus sealing the ground. If a seal is not achieved in very porous ground, bentonite slurry can be lost with risk of instability of the trenches. In order for a slurry to form a gel, the slurry must be relatively clean with suspended other soils removed.

Loss of slurry from a trench could result in the collapse of the trench. This is a major concern in very porous ground especially in karst (cavity-bearing ground) or in rubble mounds such as from abandoned and covered sea walls. In very porous ground, pre-grouting may be necessary to fill voids in the karst or the rubble mounds before trenches for diaphragm walls can be excavated using slurry. When bentonite slurry is not clean and entrains other particles of soil, a cake of mixed bentonite and other soil can form on the sides of the trench. This gooey material can remain in place when the trench is filled with concrete. The weak goo can then reduce to nearly zero any shaft friction or shaft adhesion that would otherwise be required for the wall to develop any vertical load-bearing capacity. Another problem with goo on the sides of trenches is that coarse aggregate from the concrete can penetrate the goo or the goo can penetrate the concrete between the coarse aggregates in the concrete, thereby producing a very rough surface with less than the required cover of good concrete over the steel reinforcement. Sometimes goo can be displaced by wet concrete while the concrete is poured and the goo becomes entrained, resulting in patches of soft material entrained in the concrete.

The cleanliness of slurry is important. The base of an excavated trench nearly always has some muddy material prior to concreting. That weak material should be pumped out by means of an air-lift pump threaded down between the reinforcing cage just prior to concreting. When concreting, if the slurry has not been cleaned and is contaminated with suspended particles of unwanted soil, several problems can arise. The contaminated slurry can continue to deposit sediment after the removal of an air-lift pump; then the resulting weak material remains at the bottom of the wall panel beneath the concrete thereby impairing the bearing capacity of the wall. The contaminated slurry can become entrained in the concrete and thereby weaken the concrete and impair the structural capacity of the wall or reduce cover to reinforcement. Cover to steel reinforcement is the means of preventing rusting of the steel. If contaminated slurry is not displaced it can prevent fresh concrete from covering the steel reinforcement. Reduced cover impairs the durability of the steel, and it should not be allowed. The procedure for placing concrete and displacing any slurry or sediment, is to use a tremie pipe to place the concrete. The bottom end of a tremie pipe is lowered to the bottom of a wall panel and fresh wet concrete is poured via the tremie pipe to the bottom of the wall panel. Any remaining slurry in the panel is displaced by the concrete and is sometimes mixed with the first of the concrete to arrive at the bottom of the wall panel. This mixture of slurry, sediment and concrete contaminated with entrained slurry is called laitance. Fresh concrete is continuously poured through the tremie pipe thereby displacing the laitance upwards. Fresh concrete accumulates at the bottom of the wall panel and gradually rises within the wall panel displacing the laitance above it. As fresh concrete is added, the tremie pipe is progressively lifted whilst keeping the bottom of the tremie pipe submerged below the laitance. By this means the panel is gradually filled with uncontaminated concrete and the laitance should be displaced to above the required finished top of concrete. The laitance can be trimmed off later, exposing sound concrete at the finished top level. Concreting for each wall panel should be conducted in one continuous operation; otherwise, the laitance can set at a low level and subsequent concrete arrives on top of it and does not displace it.

Therefore, when constructing diaphragm wall panels, it is important to maintain the quality of slurry. Increases in density indicate suspension of unwanted other soils; reduction in viscosity represents contamination of chemical alteration. Fundamental to control of quality is the density of the slurry. Another important test is the viscosity of the slurry. Clean bentonite skin on the sides of a trench is known to interact with concrete and improves the durability of the surface of the concrete. The presence of a skin, or its absence due to absorption into the concrete, has a major effect on the shearing forces that can be developed between a diaphragm wall and the ground. Gooey contaminated skin can be so wet that its shear strength can be less than 1kPa; then, any vertical loads on the wall panels are transmitted totally to the base with no worthwhile side friction. By contrast, having used clean slurry, sometimes the concrete becomes bonded to the ground and the shear strength of the surrounding soil can be fully developed. If a designer wishes to rely on friction between diaphragm wall panels and the ground in order to support vertical loads, then pressure grouting on either side of the panel can be considered to guarantee or enhance the shear strength of the concrete in contact with the surrounding soil.

When using cutting machines, the ground is cut into fragments; these fragments are held in suspension in the slurry for removal to the slurry treatment plant where the suspended particles are removed. When using grabs, often the soil is excavated in lumps and there is less mixing of excavated soil with the slurry. Often the slurry remains in the trench being topped up as excavation proceeds and the slurry is only changed before concreting. On site there needs to be a series of tanks or silos to hold fresh bentonite slurry and to hold used bentonite slurry. Spoil excavated with grabs is less broken up than when it is excavated with a cutting machine, and for cohesive soils, the spoil can be lumpy. The spoil is dumped at the surface of the site close to the trench because of the limited swing of the crane or gantry from which the grab is suspended. To avoid congestion on site, the spoil generally should be removed regularly. By contrast, when using a cutting machine, the slurry is pumped into and out from the trench continuously. The slurry is thoroughly mixed with excavated spoil and has to be cleaned. The pumped slurry is contained within a pipeline that leads to a slurry treatment plant. The slurry treatment is in stages: first, a grizzly (a grid of steel bars) collects pieces of hard material such as cobbles and then sieves remove sand and gravel. Removal of suspended fine soil is usually achieved by cyclone centrifuges from which relatively clean bentonite slurry is returned to the system and a thick slurry of spoil and water is piped to a filtered press. Powerful presses are used to reduce the bulk of spoil by squeezing out most of the water; the retained soft to firm clay and silt can then be carted off site for disposal. Slurry treatment plants large enough to clean slurry for cutting machines occupy an area of several dozen square metres, which mitigates against the use of cutting machines on small sites. All of the spoil is contaminated with bentonite, which has to be considered when arranging for a disposal site.

As an alternative to using bentonite, some polymers can be mixed with water as slurry during excavation. Polymer solutions seem to retain less spoil than bentonite slurries and their effects on the bond between finished concrete and the ground seem to be less troublesome, but they should be evaluated on a case-by-case basis. Some polymers are biodegradable and are sometimes preferred to bentonite in order to meet environmental concerns regarding disposal of contaminated spoil and spent slurry.

2.1.6 Hand-dug caissons

Wells have been excavated by hand for hundreds if not thousands of years. The diameters of the shafts have often been determined by the working space required for excavating and typically range from about 800mm to 1200mm. In the 1970s in Hong Kong, there was a lot of construction work for which foundation loads of the order of several hundred tonnes were required. There was a tradition of excavating wells by hand, and somebody thought of excavating shafts by hand and filling them with concrete as hand-excavated large diameter piles; these were called hand-dug caissons. The method was as follows. After digging by hand for most of a day, the worker would construct a temporary lining to the caisson, usually of plain concrete. A shutter was placed around the exposed walls of the caisson and plain concrete was placed by hand directly in contact with the exposed ground and retained by the shutter. In good ground, the excavation could be advanced by about 0.8m to 1m per day, and the perimeter was concreted on the same day. The next morning excavating was resumed for several hours then the shutter was struck and lowered ready for the next pour of concrete. Shutters were cylindrical and tapered, in the shape of the frustum of a cone, so that the shutters only required to be tapped downwards to release them for re-use without dismantling. In weak ground, timbers were driven ahead of the bottom of the caisson to restrain the soil from squeezing into the caisson. However, the method was not suitable for very soft wet ground. When obstructions such as boulders were encountered, power tools could break the obstructions.

Families were involved. Generally, the male, a husband, son or brother, worked down the caisson and the woman, a wife, mother or daughter, worked at the surface operating a bucket on a hoist to remove excavated spoil. The bucket was also used as a precarious but expedient passenger lift for the worker to go down and back up again. Accidents occurred when people fell down the open shafts, so fences were required. Life on site was precarious in those days. When rock was encountered before excavating down to a required founding level, drilling through rock was often done by hand. Drilling rock in the confined space at the bottom of caissons produced abundant respirable dust. At a typical temperature of 35 to 40 degrees, workers were reluctant to wear respirator filter masks. Within a few years, pneumoconiosis was widespread, and some 3,000 workers in Hong Kong passed away, many aged less than 35 years old. Legislation was passed to limit the use of caissons and to mandate adequate ventilation and personal protection[3]. Hand-dug caissons are now seldom used in Hong Kong.

In 1975, a local contractor tendered successfully to construct two underground railway stations and the connecting tunnels all to be built within excavations of the order of 15m to 20m deep[2]. For the tunnels, steel sheet piling was used for temporary ELS. For the two stations, at Diamond Hill and at Choi Hung, hand-dug caissons were proposed for the ELS system. For permanent walls, the secant piling method was modified and adopted, as illustrated in Figure 2.2. Primary caissons were completed with reinforced concrete instead of weak concrete. Secondary caissons were excavated between completed primary caissons, making use of two curved shutters spanning horizontally between primary caissons.

Instead of cutting the primary caissons, the workers cleaned the exposed surfaces of the primary caissons, about one quarter of the surface of a caisson on each side, and the secondary caissons were filled with reinforced concrete.

Diamond Hill Station was required to have longitudinal walls of a temporary nature, to be demolished later when the station was widened in the future according to plan. The caisson method was further adapted. Primary caissons were used to accommodate steel king posts. The secondary caissons were used to construct plain concrete arched lining on the outer side spanning horizontally between adjacent steel king posts.

2.1.7 Sprayed concrete and soil nails

In the absence of ground water, exposed faces of excavated ground can be supported to a limited extent by soil nails or by sprayed concrete, also called shotcrete. Sprayed concrete is used to bind the surface of the ground

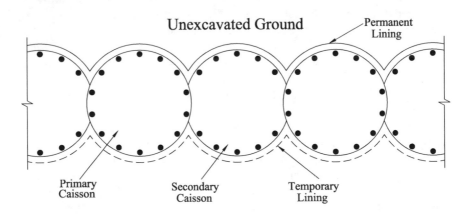

Figure 2.2 Secant hand dug caissons.

to prevent ravelling. If the surface of the ground is arched, the sprayed concrete provides some support. Soil nails are steel bars that are driven into the ground or are set in cement in a hole that has been bored in the ground. Soil nails are used to tie the ground together to a depth of a few metres. Depending on the strength of the ground, sprayed concrete and soil nails can be used as primary lateral earth support for a limited depth in soil and often to greater depths in fractured rock. The method can be used to greater depths as lagging, for example between king posts. However, below the ground water table, sprayed concrete and soil nails are less effective or may not be practical depending on the ground and groundwater conditions.

2.1.8 Selection of type of wall

Diaphragm walls are often used for temporary and permanent walls. Diaphragm walls are generally constructed in panels about 6m long such that a wall 60m long would have only nine or ten joints, whereas a wall formed with 1m diameter piles would have 59 or 60 joints and more opportunities for leakage. Vertical joints between adjacent diaphragm wall panels are usually formed as stepped joints, and often joints incorporate PVC water stops. In the 1980s, there was a practice of overlapping horizontal steel reinforcement between adjacent panels, but this practice had a high risk of problems such as misalignment of the steel, wrapping of the water bar and inclusion of soil in the joints. This type of joint is now seldom used if at all. Even with a lot of care, joints can leak. Also, sound concrete is slightly permeable and transmits tiny amounts of water from the ground. Diaphragm walls can be used on their own as permanent walls where slight seepage can be tolerated, say in underground car parks. Additives can be mixed with concrete and can almost eliminate transpiration of ground water for dust-dry conditions where needed. Diaphragm walls are cast against an excavated surface of soil and are usually rough. Often a screen wall is built close to diaphragm walls with a gap between the screen wall and the diaphragm wall for vapour control, for collection of seepage water and for architectural finishes. Walls built from piles are not often used for permanent walls because of the concerns about seepage of ground water. Examples of pile walls are some of the first metro stations beneath Nathan Road in Hong Kong where the walls are exposed but seldom noticed because of the many illuminated advertising panels mounted on the walls. Pile walls have a rough appearance. Permanently exposed pile walls are few. One basement carpark ramp at JEM, Jurong, Singapore, even has white painted exposed pile walls.

2.1.9 Bracing

For excavations in soil and less deep than about 5m, strong retaining walls can be used to support the ground without bracing by acting as cantilevered earth retaining walls. However, for deeper excavations, bracing is generally needed to support the retaining walls laterally. For excavations of limited lateral extent, bracing is commonly provided by struts spanning from one side of the excavation to the other. Bracing can be used temporarily while excavating, or the permanent structure, in whole or in part, can be built by the top-down method as described in Section 2.2.2. For very wide excavations, long horizontal strutting may be too expensive or too flexible and inclined strutting might be adopted; support may be achieved by forming abutting berms of unexcavated ground; or tieback anchors may be installed through the walls into the surrounding ground. For stiff ground, soil nails may be used instead of anchors.

In order to provide headroom for a mechanical excavating plant, horizontal bracing is usually installed at successive depths of 2.5m to 4m. Horizontal bracing often comprises simple steel struts made from single rolled steel sections, or it may be built up from two or more steel sections strapped together. Sometimes steel tubing is used, or struts may be fabricated according to a special design. In order to make good use of the capacity of simple steel struts, secondary bracing is used to prevent the struts from buckling. Vertical restraint is often provided by driving vertical posts prior to excavation and tying the bracing to them. Lateral bracing is often provided using steel ties or beams. Steel tubes and fabricated struts may have sufficient stiffness that bracing is not required. Steel strutting has the benefits of quick erection, easy removal and possible re-use. Steel bracing is very common, and some companies offer a hiring service for steel bracing. The capacity of struts is often greater than a single element of the wall. Horizontal beams, called waling, are used to span two or more wall units such as king posts, several contiguous piles or secant piles. Steel beams are often used as waling; sometimes waling is formed from reinforced concrete beams cast against the walls.

Bracing between two parallel walls is straightforward with waling, struts and bracing set up in a simple orthogonal three-dimensional grid. Sometimes walls are not straight or not parallel, and corners are not always at right angles. In such cases, the layout of the steelwork may include diagonal or inclined struts, and adjacent struts and waling may be combined to form trusses or space frames. Corners are often braced with diagonal strutting for which the inclined strutting has to be restrained laterally at the wall. Special consideration is required at re-entrant corners where adjacent waling may have to be joined to prevent the abutting walls from separating. A common practice is to provide a capping beam on the top of the walls linking the elements of the walls together before bulk excavation is carried out.

Inclined strutting may be considered for use when the width of the excavation is large. One method involves excavating with temporary slopes forming berms against the walls and providing space for the erection of inclined struts. The tops of the inclined struts must be restrained to prevent sliding up the walls. At the bottom, a foundation is required. There are several solutions. One is to cast a reaction block of concrete. Another is to construct part of a base slab for the inclined struts to bear against. A third method is to provide piles at the toes of the struts. When further excavation takes place, the piles become exposed and can be designed to act as cantilevered columns, which then require further support, say, by extending the inclined struts to another row of piles at a lower level.

Tieback anchors are formed by drilling through the walls into the ground outside the excavation area for a distance far enough to achieve reliable anchorage. High strength steel bars or bundles of high tensile steel cables are threaded through the walls and grouted into the ground. On the inside face of the wall, a bearing plate is added, and the steel bars or steels cables are pre-loaded against the wall, thereby tying the wall back to the ground. Problems occur when the anchorage length is too short and yields, the grouting is incomplete or the steel is exposed and corrodes. In tropical climates such as Hong Kong, corrosion can be quick. Stress-induced corrosion can be faster. Steel bars can accommodate some rusting (e.g. 1mm pitting on a 50mm diameter bar is not large, and steel bars can be galvanised for protection). Steel strand comprises wire of small diameter, say 3mm, and 1mm of corrosion severely reduces the area of cross section of a wire. Furthermore, corrosion can be faster with high-stressed strands than with some of the lower-stressed steel reinforcement bars. In Hong Kong, there were incidents of anchor heads failing due to corrosion, and one incident of an anchor head flying off into a schoolyard, fortunately with no fatalities. This event in Hong Kong resulted in specific regulations controlling the use of tieback anchors[4] such that they are now seldom used and those that are used have double protection against corrosion, as well as regular monitoring.

Permanent bracing for lateral earth support makes use of parts of the permanent structure, usually reinforced concrete, built in stages as the excavation is advanced. Sometimes steel bracing is incorporated into the permanent works, but it is more common to use reinforced concrete. This can be achieved by constructing beams or parts of slabs that span across the excavation. This is the top-down sequence of construction and is described in Section 2.2.2. The procedure is to construct parts of the permanent structure by casting reinforced concrete beams or slabs on blinding on the excavated ground with no need for formwork. Steel bracing is often pre-loaded by jacks and locked off, but concrete shrinks as it cures and subsequently creeps. Also, concrete takes time to cure and gain in strength, whereas steel is ready to use immediately. However, steel bracing is quite costly, and using some of the permanent structure for bracing saves funds for the temporary bracing. Building the permanent structure piecemeal can increase the cost of the permanent structure and takes more effort to design. When given a reference design with temporary bracing, construction companies often offer alternative designs with permanent bracing, saving in cost and time. For example, Diamond Hill Station in Hong Kong[2], mentioned earlier, was built according to the contractor's design adopting a top-down method whereby both the roof and the concourse floors were cast on blinding during the excavation and steel bracing was not used. At the same time, Choi Hung Station was designed and built with main beams for the roof that were constructed on blinding and spanning from wall to wall at the locations of the columns. Spaces between main beams permitted plentiful access when working below the level of the roof; they were later filled with pre-tensioned Tee-beams and cast-in-situ concrete topping. The whole roof was not constructed early because the station is wide, and the weight of the whole roof would have exceeded the bearing capacity of the temporary pier foundations beneath the columns prior to their connection to the base slab. Ultimately, the uplift beneath the base slab balanced the weight of the roof and the backfill.

For very deep excavations in soft ground, bracing with many levels of struts and thick diaphragm walls can be uneconomical. For multiple strutting, the maximum deflection of the walls is usually below the excavation level and the maximum displacement increases as the excavation is lowered. Therefore, installing bracing before excavation greatly reduces the inward displacements of walls at depth. Buried struts have been formed using jet grout before excavating, creating overlapping piles spanning from wall to wall at specified depths. Individual rows of piles can form buried struts, and a two-dimensional array of JGPs can form a slab between the external retaining walls, bracing them apart from the moment that the excavating starts. Jet grout slabs can be permanently located below formation level or temporarily above the final formation level to be removed while excavating. Both temporary and permanent jet grout slabs have been used at the same time, as identified in the discussion about the collapse of the Nicoll Highway in Section 5.2.3.

2.2 Sequences of construction

2.2.1 Bottom-up sequence of construction

Section 2.1.9 describes the use of permanent bracing, the top-down sequence of construction for deep excavations. The traditional sequence of construction is bottom up. By this method, the excavation is completed to the final level and then the permanent structure is built from the bottom up. The bottom-up sequence of construction is illustrated in Figure 2.3. Temporary ELS is usually adopted. Retaining walls are usually built from the ground surface, and soil excavation commences at the ground level. Temporary bracing is installed stage by stage as excavating

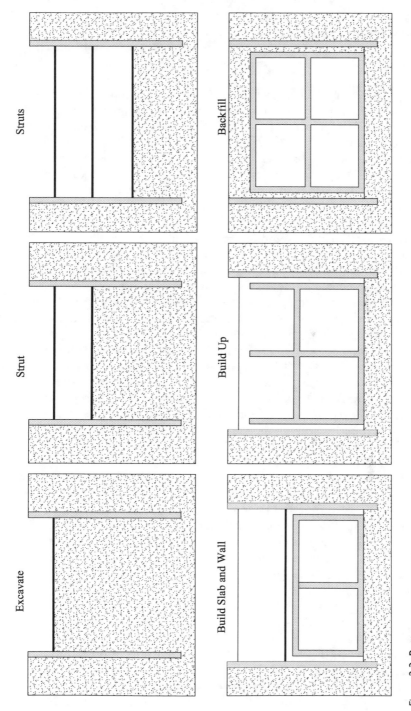

Figure 2.3 Bottom-up sequence.

progresses down to the final level. From the final level, the permanent structure is built, bottom up, independent of the temporary ELS. The ELS is dismantled in stages as the permanent structure is built, and the space between the permanent structure and the sides of the excavation is backfilled.

A gap is required between the temporary walls and the permanent walls for formwork for constructing the permanent walls. After the construction of each lift of permanent walls, the formwork can be removed, durable waterproofing materials can be applied to the outside of the permanent walls and the gap between the temporary and permanent walls can be backfilled. Where space is critically limited, the total thickness of temporary walls, gap and permanent walls may be wasteful. In such cases, one wall (e.g. a diaphragm wall) can be used for both a temporary wall with temporary bracing and a permanent wall; then, the other permanent internal walls, columns and floors are built bottom up.

During bottom-up construction, bracing is removed in stages as intermediate floors that permanently brace the walls are constructed. Levels for bracing should allow for the unobstructed construction of the permanent structure in stages; otherwise, construction has to be modified into more stages. Alternatively, a permanent structure can be built around the temporary bracing, omitting the need to fill in around the bracing later. Sometimes, more temporary bracing is added so that any obstructing bracing can be removed. The design for permanent works using the bottom-up method of construction is straightforward since the structure can be built and backfilled without considering temporary stages of construction and the locked-in deformations and stresses that result from top-down construction. By keeping the permanent structure independent of the temporary ELS, the construction of the two can be separated and, if required, built under separate contracts.

Fifty years ago, when deep excavations and diaphragm walling were not common (e.g. projects such as the New World Hotel in Hong Kong in 1975), excavation and temporary works were completed first, and time could pass before a permanent structure was built bottom up. The New World Hotel was the first project in Hong Kong to use diaphragm walls (called ICOS walls at the time; the ICOS company was a pioneer in diaphragm walling) and tieback anchors permitting unobstructed construction of the six-floor basement.

Bottom-up construction occupies a lot of the site area, because in most cases nearly all of the site is excavated. Any unexcavated ground generally is only around the periphery. Site area for offices, materials and equipment can be in short supply. Cranes are needed to lift excavated spoil and to lower plant and construction materials. For large excavations, cranes located around the outside of the excavated area may not be able to reach across the site, and temporary decks covering part of the excavation may be required for cranes to reach the working areas. For some very large excavation, roadways with ramps are adopted so that trucks can drive up and down to remove excavated materials and deliver plant and materials for construction. For narrower excavations, temporary gantries and hoists are sometimes built to span the excavation for purposes of accessing the whole area for the removal of spoil and waste materials and to lower materials.

For many basements, the whole of the site is excavated, and temporary decking is necessary to provide a work platform at ground level. In some cases, where the site area is small, temporary multi-storied structures are built for offices, storage and processing plants (e.g. a slurry treatment plant), and ground-level decking is a priority for vehicular access.

2.2.2 Top-down sequence of construction

As Section 2.1.9 describes, in top-down construction parts of the permanent underground structure are built concurrently as the excavation is carried out in top-down stages. The sequence of construction is illustrated in Figure 2.4. First, the outer walls are constructed from the ground surface. Excavation is carried out in stages progressing to each floor level. Permanent floors are constructed at each stage of excavation. The permanent floors extend across the excavated area and act as bracing to the external walls. Often there is no need for temporary bracing. Usually the outer walls are used as both temporary and permanent walls. Sometimes, temporary walls are supplemented by inner walls, and the two walls act compositely in the long term.

For building works, the top-down method offers the advantage of an early start on constructing the aboveground structures and simultaneous excavation and construction of the underground permanent and aboveground structures. The method requires that foundations be built early, and since these necessarily extend below the base of the excavation, plunge columns from ground level are required to pass through the depth of the proposed excavation to support the superstructure until permanent columns are completed. For convenience, plunge columns can be located at the permanent columns and be included within them. The Sands Hotel and Casino at Macau was built with this sequence, which shortened construction by an estimated six months compared to not commencing the aboveground structure until after the basement had been built from the bottom up.

During top-down construction, in order to provide enough of the permanent structure to brace the walls, floors are sometimes cast all at once, making use of permanent openings for access during construction. At other times, temporary openings are used for wider or direct access during construction and the permanent structure is completed later. Sometimes permanent beams

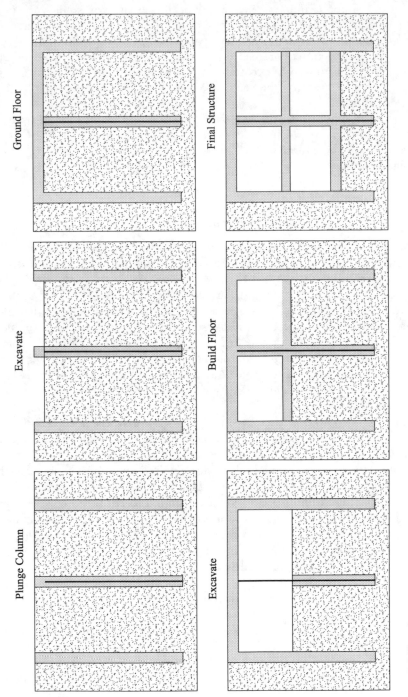

Plunge Column

Excavate

Ground Floor

Excavate

Build Floor

Final Structure

Figure 2.4 Top-down sequence.

are constructed for use as temporary bracing and are later incorporated into the permanent structure.

Top-down construction has advantages of using very little or no temporary bracing. Commencing with the ground level floor, the top-down method provides a large and stable decking. However, the design for permanent structures built top-down including temporary stages is more complicated than the design for structures that are built bottom up independent of the temporary works.

2.2.3 Hybrid sequences of construction

Sometimes the site or the project requirements lead to the adoption of hybrid schemes using partially top-down and partially bottom-up sequences of construction. For example, South Island underground railway station is located alongside and below part of Admiralty Station in Hong Kong[5]. The top-down method was adopted for the first three floors when excavating in soil. When bedrock was reached, ELS was no longer needed and open excavation went ahead with increased headroom and large equipment could be used. Within the rock, the permanent structure was completed by the bottom-up method of construction.

2.2.4 Advantages and disadvantages of the methods

The bottom-up method of construction is simpler to design, and the excavation and temporary earth lateral support are completed before permanent structures are built. The permanent structure, if kept separate from temporary works, has no locked-in deformation and stresses due to the sequential loading during excavation. By building the permanent structure inside a braced excavation, it can be surrounded by waterproofing. Waterproofing can be placed beneath the base slab and extended up the outside of the permanent walls and to the ground surface or over the roof as appropriate. For the bottom-up method, when the width of the excavation exceeds the reach of cranes, a deck at ground level is needed for cranes to hoist spoil and for other activities. Roadways are needed across the site for vehicular access for removal of spoil and for delivery of materials and plant. All the permanent concrete can be cast against formwork with the required quality of finish. In most cases, joints between concrete structures can be provided with normal laps for steel reinforcement avoiding the use of couplers on short bars.

The top-down method makes use of some of the permanent structure for temporary earth lateral support thereby saving the use of temporary walls and temporary bracing. On large congested sites with little or no other work areas, top-down construction can include substantial areas of permanent floor that provide clean level platforms for use as work areas.

The top-down method allows simultaneous construction of above-ground structures. Early construction of floors also provides cover or partial cover to the site, which is useful protection from the weather. Rainfall generally halts excavation work, and shelter from the rain increases productivity. Concrete floors can be constructed on blinding placed on the excavated ground surface. Formwork is not needed. Concrete cast on blinding, or on plywood sheets laid on the excavated surface or on the blinding, provides a smooth finish to the soffit of the concrete and is far less expensive than formwork that is required when building from the bottom up.

Top-down construction can often be cheaper or quicker to construct than the bottom-up method, and the top-down method is popular amongst contractors when competitively bidding for design and construct contracts. However, the design is more complicated and more expensive, and the construction sequence is more complicated than when using the bottom-up method.

2.3 Basic theory

Soil includes clay, silt, sand, gravel, as well as cobbles and mixtures of these basic types. The particle sizes of soils range from the microscopic crystals of clay that require a scanning electron microscope to be seen to large fragments up to 200mm diameter, which is the technical maximum size of cobbles. However, soil can also include embedded boulders and a variety of artefacts. Just as particle sizes of soil vary widely, the mechanical properties of soil also range widely from mud as soft as toothpaste to hard saprolite (decomposed rock), which is almost impossible to drive a steel bar into by hand using a hammer, yet it slakes (it falls apart in water). Some soils are very permeable and below ground water-table rapid rates of inflow would flood the excavation, but some clays are so impermeable that they can be used as impervious liners to dams and reservoirs.

Characterisation of the types of soil that are present at the site, their types, disposition and mechanical properties is fundamental to planning, design and construction of deep excavations. It is important that locations, extent, strength, stiffness and permeability of soils are quantified and appropriate parameters are determined for purposes of design and planning the construction of deep excavations. In simple terms, strength of the ground is fundamental to determining the stability of the sides of the excavation and the loading on the ELS system supporting the sides of the excavation to keep them stable. Stiffness is related to how much the ground will move and whether such movements however small will damage any property such as nearby structures and utilities including gas mains and water mains. Gas and water mains require protection because if they are ruptured, they have a high consequence of fire and flooding. Permeability affects how much ground water would seep into the excavation when excavating below the ground water table. Seepage of ground water into an excavation must be limited to provide safe working conditions. Seepage also must be limited to prevent subsidence of the surrounding areas due to loss of ground water. Arising from excavating and removing soil, confining pressures in the adjacent ground generally are reduced and the permeability of the soil controls the rate at which the soil can swell and absorb water. When soils swell, they generally soften. The degree by which the confining pressure is reduced when the strength of soil is reduced should be taken into account when designing deep excavations.

Before explaining how to determine such properties, an explanation of how soils behave is warranted. All soils comprise free particles, with spaces (pores) between them that may be full of air when the soil is dry or full of water when the soil is saturated and may be partially full of air and water where it is partially saturated. The particulate nature is easy to appreciate for a handful of dry sand that can run through one's fingers. It is less easy to appreciate for clay, which has such small particles that we cannot see them without a microscope. For hard clays, it is possible measure a tensile strength with a quick test, but by adding water to hard clays and to softer clays, the material softens and eventually the particles disassociate. Adding a lot of water to softened clay and giving it a shake results in a suspension of fine particles that takes a long time to settle as sediment. For engineering purposes, key properties for all soils are strength and stiffness. Strength and stiffness can be measured by tests. However, these properties can change with time. Strength and stiffness of a given type of soil depend on the extent of pores in the soil, how much water, or air, is contained in the pores between the particles and how quickly the water or air content can change. For example, adding water, or allowing water to seep into a soil, as can happen for a deep excavation, generally weakens a soil.

Scientific principles were first applied to the behaviour of soil in the 1700s by the French military engineer, Coulomb[6] and were developed later in the 1800s by the scientist Rankine[7]. Systematic study of the mechanical behaviour of soils commenced in the 1930s in the United States with fundamental publications on soil mechanics by Casagrande[8], fundamental textbooks by Taylor[9] and classical theory by Terzaghi[10]. Some of the early theoretical concepts have been modified or updated. Holtz et al. [11] have authored a more recent textbook with an introduction to soil mechanics. Soil mechanics is now often referred to as geotechnical engineering. A textbook by Atkinson et al. [12] is a good introduction to the mechanics of soils, including complicated constitutive models such the Critical State Theory for soil mechanics.

2.3.1 Strength of soil

Open-textured soils, such as clean silt, sand and gravel are very permeable to air and water. If permeable soils are compressed, air or water can escape quickly, the particles become pressed closer together and there is more resistance to shearing. When confining pressure is reduced, air and/ or water can enter as the soil particles relax and the shearing resistance is reduced. This dependence of shearing resistance to confining pressure is called frictional behaviour and soils that drain in this way have frictional properties. Shear stress, τ, is related to effective confining stress, σ', by a coefficient of friction. Friction is defined as an angle, ϕ, and the relationship between τ and σ is given in Equation (2.1)

$$\tau = \sigma \tan \phi' \qquad (2.1)$$

Friction angles for clean sand can vary between about 28 degrees and 42 degrees depending on the mineral composition and on the roughness and the angularity of the particles. Clean sand and gravel have high permeability are quick to drain. At the other end of the range, clay has very low permeability and is very slow to drain. Nevertheless, when time is allowed for drainage to take place, clay also exhibits frictional properties and values, for the angle of friction for clay is typically in the range of about 16 to 23 degrees.

It may be noted that some soils, when tested, do not exhibit an exact relationship between τ and σ; the trend usually is a curved envelope with a slightly higher value of ϕ at low stress levels and a lower value of ϕ at higher stress levels as sketched in Figure 2.5. A common practice, as an approximation, is to construct a tangent to the curve at the range of engineering stress levels. The tangent makes a small cohesive intercept called c' in the range of, say, 2 kPa to 5 kPa and an average value for ϕ that applies at that range of stress. The two parameters are reported, c' and ϕ, as drained strength parameters. The apparent cohesion is only a result of the curved strength envelope and is not a true cohesion that sticks the particles together. A curved strength envelope is usually associated with soils dilating when sheared due to interlocking soil particles at higher stress levels.

2.3.1.1 Why clay appears to have cohesion

Why does clay exhibit cohesion, one might ask? Low permeability soil, such as clay, takes a long time to drain. When a specimen of clay is pulled in tension or sheared, a resistance is experienced, and a shear strength can be measured. When some clay is stirred up in water it disassociates, showing that when water is added there is no intrinsic shear strength. The tensile strength and shear strength occur because the tension or shearing induces negative pressure in the water in pores of the clay particles. At a large scale, and when loading is of relatively short duration, say during the excavation of a deep basement, and no significant drainage

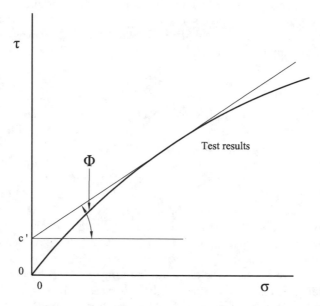

Figure 2.5 Characteristic curved drained strength envelope.

occurs because the soil is of very low permeability and saturated or almost saturated with water, the shearing strength of the soil is unaffected regardless of the confining pressure. The soil exhibits a constant "cohesive" shear strength, which is commonly given the symbol "c_u". The suffix "u" denotes that the strength applies as long as the soil remains undrained. If and when drainage takes place, the strength alters according to the change in confining pressure as described above.

   This fundamental behaviour of low permeability soils is sometimes difficult to comprehend. Lumps of clay feel cohesive like plasticine or as strong as very dry cheese requiring an appreciable force to break. However, the strength of a group of particles of soil depends on the contact pressure between the particles of soil and not on the water present in the spaces between the particles. Clay is compressible; one can squeeze the water out of it. An increase in compression of a group of clay particles increases the strength of the particles and reduces the volume. If the spaces between particles are occupied by water and there is no time for the water to be squeezed out, the overall volume of the particles does not decrease, and their strength does not increase. The increase in confining pressure is balanced by an increase in the water pressure and not an increase in the stresses on the soil particles until drainage occurs. This phenomenon is described by the Principle of Effective Stress. Total stress is applied to a mass of soil. The effective stress is the amount of stress supported by the grains of soil, the soil skeleton and the effective stress plus any stress in the water in the pores of the soil adds up to balance the applied total stress. Effective stress equals total stress minus pore water pressure.

2.3.1.2 Partially saturated soil

Determination of the strength of soil becomes more complicated when air is entrained in fine-grained soils because the air is compressible and does not have to escape or enter the soil, as water does, for the volume of the soil to change. This results in an intermediate condition; when, for example, the confining pressure is increased, some of the air voids get smaller and the soil particles move a little closer together. However, the amount by which the volume of soil decreases is less than when the water is also expelled. Therefore, the increase in the strength of the soil is less than if the soil is fully drained, the frictional effect is reduced and the measured value of ϕ is smaller. Because the soil is not fully, but only partially, drained, the cohesive effect is still there but less than if the soil is fully undrained. The study of partially saturated soil is specialised, and for many deep excavations the zones of partially saturated soil, if any, are above the ground water table and are often small.

Many natural soils contain a mixture of soils with a range of particle sizes, and often there are mixtures of coarse-grained soil and fine-grained soil. The fundamental concern for deep excavations is whether the soils drain significantly during the excavation process. Lack of drainage is controlled by the percentage of the content of fines, especially the finest size, clay. The proportion of fine-grained soil, clay and silt to coarse-grained soil, sand and gravel, is germane. Small quantities of fine-grained soil can be located within the pores between coarser particles. If the volume of fine-grained soils is less than the pores between the coarse-grained soil particles, there will be pathways allowing water or air to move through the mass with little resistance, and the soil mixture can drain quickly. However if the fine-grained soils completely fill the pores between the coarser particles, water or air would have to pass through the pores between the fine-grained particles and, being unable to pass through the coarse-grained particles, the permeability of the mixed soil would be less than the permeability of the fine-grained soils themselves and the mixed soil would be slower to drain than the fine-grained component. As a guide, in a well-graded soil, 35% fines content by weight would result in very low permeability of the mass.

2.3.1.3 Saprolite

An exception is saprolite, a special type of soil. Saprolites are formed in situ by the decomposition of rock. Grade V decomposition is when rock has decomposed to soil, yet the fabric of the parent rock is still present. For example, granite contains quartz, feldspar and biotite in the main. Completely Decomposed Granite (CDG) comprises quartz sand with packets of clay, mostly kaolinite, at the locations of the feldspar crystals and mica where the biotite has decomposed. The kaolinite tends to be in the form of small packets of sub-parallel clay crystals. Because of this microstructure, water can flow between the small packets of clay crystals more freely than if the kaolinite were without structure or if it were uniform. Moreover, contacts between quartz particles have been dissolved creating tiny passageways through which water could flow. Therefore, the tightly packed crystals of quartz and the enclosed packets of clay have an open structure and are conductive of water. This combined effect is that CDG is relatively permeable and during a deep excavation generally behaves as if it were totally drained, even though the grading of particle sizes with 35% or more fine-grained soil would result in very low permeability if the particles were uniformly mixed together with no residual fabric. Likewise, it is noted that fissures in stiff clays, another type of fabric, further complicate the issue. Intact stiff clays have very low permeability, but the fissures can be conductive. If the fissures are interconnected, then the mass conductivity can be considerably enhanced. The structure of the in-situ soil affects its properties such that the mass properties differ from those inferred from the grading analysis of particle size and as inferred, assuming that the particles are uniformly mixed.

2.3.2 Measurement of strength of soil

Several tests can measure the strength of soil. Tests for strength can be carried out in the field or on samples that have been retrieved from the field and are tested in a laboratory. The tests can be of short duration in an attempt to measure undrained strength or sustained to allow drainage to take place in an attempt to measure drained strength. Care should be taken to determine whether the soil is undisturbed or has been disturbed before being tested. Many years ago, at a meeting of the Hong Kong Institution of Engineers on the topic of strength of soil, a luminary of the industry made a comment along the lines of *"We all know that at a particular point in time a piece of soil has a unique strength."* With so many impressionable young engineers present and eager to learn from "The Masters," I could not let the matter pass. I replied that the comment should be understood in the light that one should consider whether the soil is drained or undrained during testing and that undrained tests using different pieces of apparatus such as a triaxial machine, a shear box, a simple shear apparatus or a vane shear apparatus result in different values of strength, certainly for clays the rate of loading affects the strength and, for natural soils there is often anisotropy. Strength of soil does vary according to the type of test and the rate of straining. Strength of soil can also be inferred from other tests, such as Standard Penetration Test (SPT) and Cone Penetration Test (CPT) and Atterberg Limits as described later in Section 2.3.2.3, and there are correlations in use that relate strength to other tests that are not intended to measure strength.

2.3.2.1 Disturbance

Several tests are commonly used to measure the strength of soil in the field; an objective is to test soil that has not been disturbed or subjected to the minimum disturbance. Disturbance affects the strength of many soils. The most alarming effect is for sensitive clays. Sensitive clays are mostly clay soil that was laid down by sedimentation in a salty marine environment but subsequently the salt has been leached by percolating fresh water. There are several effects of disturbance on the strength of soil. The most spectacular example is "quick clay" such as is found at Drammen in Norway, which, when undisturbed, is firm enough for people to stand on without sinking but when remoulded, the soil becomes a suspension with the consistency of hot chocolate. Quick clays occur because when they were deposited in salt water, they were flocculated in salt water. The clay particles settle and sediment from suspension and consolidate on the seabed with a very open texture. It is considered that in salty water the clay particles become charged so that flat sides of particles repel each other but edges of plates of the opposite charge are attracted to flat sides. Changes in sea level can result in marine sediments becoming exposed on dry land and subjected to infiltration of fresh rainwater or groundwater. When the salt is removed by fresh water the electrical charge changes; when disturbed, the house of cards type of structure of the flocculated clay collapses, there is free water between the loose particles of clay and the strength drops accordingly until the clay is consolidated. Many clays, called sensitive clays, exhibit such behaviour and the sensitivity can be determined by recording the peak strength for undisturbed soil and a much less residual or remoulded strength by continuing the test for large deformation or by remoulding the soil. The sensitivity is the peak strength divided by the residual strength. Marine clays beneath the seabed in Hong Kong and Singapore have sensitivity of the order of 3 to 4.

Sand also can be disturbed. Sand that is dense dilates when disturbed; the particles become further apart, and the overall strength is reduced. Sand that is initially loose tends to become more compact when disturbed, and the overall strength increases. The stability of a deep excavation depends on the strength of the soil that is adjacent to an area that is to be excavated and is to be retained by ELS, and soil within the excavated area that comprises the base of the excavation at each stage of excavation is initially undisturbed; for a safe excavation, the soil should remain largely undisturbed. For sensitive or quick clays, it is important to ensure that disturbance caused by the excavation work is not enough to exceed the peak strength and induce collapse. Whereas it is standard practice to maintain a factor of safety to prevent failure, sensitive clays should not be over-stressed locally, and the margin should be big enough to ensure that the clay does not creep at a loading that is less than the peak load but reaches a large enough strain to disturb the brittle loose structure of the quick clay and induce a collapse. I am not aware of any deep excavations carried out in quick clays where the consequence of disturbance amounted to a total loss of strength and special care was required.

2.3.2.2 Standards for testing soil

There are standards for testing soil including the strength of soil. The British Standard for testing soil is BS 1377[13]. In Hong Kong, the Government has published a guide to site investigation and includes methods of investigation and testing soil[14]. Field tests to measure the strength of soil include the vane shear test, which comprises a small vane with four blades at right angles pressed into the ground at the base of a borehole. The vane is turned and the torque that is applied and the angle of rotation are recorded until the soil is sheared with a vertical cylindrical plug of soil that is trapped between the four blades. The shear strength of the soil is developed around the cylindrical perimeter plus the circular top and bottom of the plug of soil. Peak strength is determined by measuring the maximum torque that was applied to the vane. The peak strength is recorded and usually the vane is twisted through several rotations to measure the residual shear strength as well. Other tests conducted down boreholes are the flat plate dilatometer and the pressuremeter. The flat plate dilatometer is a pair of flat plates initially close together like a flat blade that is pushed into the ground at the base of a borehole. The plates are pushed apart hydraulically, and the pressure is recorded. Determining the strength of the soil from the measured pressure requires skilled interpretation. The standard Menard pressuremeter, named after its inventor Louis Menard[15], comprises an inflatable packer that is lowered down a borehole and is expanded by applying pressure; the volumetric expansion is recorded along with the increases in pressure.

The Menard pressuremeter is placed in an open borehole where the sides of the borehole have been exposed and might have been disturbed; the rubber membrane around the pressuremeter has to be bedded onto the exposed face of the soil. These are undesirable factors; a self-boring pressuremeter[16] was devised Peter Wroth[17] and John Hughes[18]. It has a cutting tool at the base so that the test probe can be sunk into the soil at the base of a borehole with the intention of minimising the disturbance of the soil. When conducting a test, the device is inflated, the pressure is recorded and the radial displacement is measured. For both types of pressuremeter, skill is required in interpreting the results in order to determine shear strength. Skill is needed because shear strength is developed progressively starting at the perimeter of the cavity formed by the borehole and as the pressuremeter is expanded soil reaches its peak strength at progressively increasing radii. The dilatometer and pressuremeters do not measure strength directly. Details about testing soils with pressuremeters are provided by Mair et al [19].

Data from probes that are driven or pushed into the ground can also be interpreted, indirectly, to measure the strength of the soil. The SPT comprises a conical tip on a series of rods that are driven into the ground at intervals from the base of a borehole. A standard hammer of 63.5 kg weight is dropped a standard height of 760mm, and the number of blows required to drive the probe for three successive distances of 150mm into the ground is recorded. The blow count for the second and third distances of 150mm are added and reported as the SPT N number. The number of blows for the first 150mm penetration is not included in N value because the soil might have been disturbed by the boring of the hole. The test is a measure of the resistance of the soil to penetration by the conical tip; it is not a direct measurement of the strength of the soil. There are several correlations of the N value with strength. Another type of probe is the CPT. In this test, a probe comprising an instrumented conical tip and a short, instrumented shaft of the same diameter is pushed into the ground. The force on the tip and the developed shearing force on the following shaft are measured as the CPT device is pushed into the ground. Some CPT devices include a porous gauge to measure the pore water pressure developed either on the shaft or on the tip. Skilled interpretation is required to determine the material type, strength and stiffness of soils from the test results. A definitive account of such methods is given by Robertson and Campanella[20]. A general guide to the interpretation of CPTs is given by Mayne[21].

2.3.2.3 Tests in a laboratory

When samples of soil have been retrieved from the ground they can be sent to a laboratory where a variety of types of testing can be carried out under controlled conditions. A description of laboratory tests is given by Head [22]. Unfortunately, there is a risk that specimens may have been disturbed by the sampling, transportation and by trimming in the laboratory. When evaluating the results from tests, one should consider whether any of the specimens have been disturbed; this is often done by correlating the results from different types of tests. Sampling disturbance can be reduced by several techniques such as using triple tube coring tools for taking the samples, thin walled samplers for very soft soils and foam as a flushing medium while boring. Of course, care should be taken in handling, transporting and preparation of the specimens for testing. A number of tests can be carried out in a laboratory that cannot be carried out in the field. A merit of testing in the laboratory is that drained tests can be conducted on small specimens taking several days or more, whereas in the field tests tend to affect a much larger mass of the ground and can take a much longer time.

Basic strength tests that are carried out in a laboratory have been mentioned above. A commonly used test is the triaxial test, whereby a cylindrical specimen is wrapped in an impervious membrane. The specimen is surrounded by fluid in a cell that can be set at a prescribed confining pressure. A vertical ram can be used to load the specimen axially. Some tests are conducted by increasing the axial load until the sample reaches a peak load or until it fails. Such tests are called strain-controlled tests. Other tests are conducted by axially compressing a specimen by pressing the vertical ram downwards whilst recording the load applied to the specimen by the ram. Such tests are called strain-controlled tests. Strain-controlled tests are useful when soils exhibit a distinct peak stress beyond which continued increases in strain results in decreasing axial load. Strain controlled tests can be used to record diminishing axial load as a specimen is strained beyond a peak load condition, whereas stress-controlled tests would result in the collapse of a specimen and no measurements of the diminishing strength post peak.

In the standard triaxial apparatus, a specimen can be sealed within a rubber membrane and maintained at a constant water content in an undrained test or it can be allowed to drain in a drained test. For soils with low permeability, drainage in the field can take a long time, such as weeks, months or years. During the time for excavating a deep excavation, typically many months, low permeability soils such as clayey soils, can remain undrained or only drain partially, while soils with a higher permeability such as sandy soils can drain more completely. In a triaxial cell, the specimen sits on a porous base through which the drainage can be controlled. Commonly at the start of a test, water under pressure is applied to a specimen via the porous base in order to saturate the specimen with water. This measure is required because although the soil is often saturated in the field, as the result of sampling, transportation and preparation it sometimes loses moisture and the specimen is only partially saturated with water in its pores. Complete saturation with water is then necessary to replicate the condition of soil in the field when the test requires a saturated specimen.

When conducting an undrained test on a saturated specimen, the strength is independent of the confining pressure and only one test is carried out on a sample. Drainage results in the strength of soil varying depending on the confining effective pressure. In order to determine the drained strength parameters, it is necessary to measure strength of the same soil at more than one confining pressure. It is commonly done by taking three specimens from the same sample and testing each to failure at different confining pressures. There is concern that three specimens, each typically

150mm high, might comprise different soil, especially natural soils that have been deposited in relatively thin layers; an alternative procedure is to test the same specimen at three successively increasing confining pressures in a multi-stage test. For this procedure, a concern is that the fabric of the specimen might be altered or disturbed during the first or second stages and that the performance of the specimen during the second and third stages would not reflect the performance of an undisturbed specimen. There is a risk that the disturbed strength and not the undisturbed strength is measured in the second or third stage of a multi-stage test. Therefore, caution should be exercised when using multi-stage tests. The test results should be carefully examined for any evidence of damage or disturbance to the specimen. For example, if the shear stress shows a distinct peak strength and a reduced strength when the specimen is further strained, it can be suspected that the structure of the specimen has been disturbed or altered and that during subsequent stages under higher confining pressures the specimen may yield at lower strength than if it had not been subjected to a previous stage. Further details about triaxial testing are given by Lade[23].

Note that the conventional triaxial test is not a truly triaxial test because the cell pressure is all-round. The correct description of the test is "axial compression of a cylindrical sample." Triaxial means "applying on three axes." A true triaxial test would involve loading in three orthogonal directions independently. Various triaxial testing machines have been developed as research tools but they are not used commercially to any significant extent.

An early type of testing machine that is still in use is a shear box. A specimen is trimmed to fit inside a stiff metal box. The top half of the box contains a piston that can apply a vertical load to the top of a specimen. The two halves of the box can slide horizontally, one over the other, and induce a horizontal rupture in the soil specimen. The forces on the boxes are measured and used to determine the shearing stress on the horizontal plane of the soil specimen. The tests can be repeated with fresh specimens at different vertical confining pressures. Criticism of the basic apparatus is that the specimen can drain in an uncontrolled fashion because the apparatus is not sealed. Also, a narrow rupture is induced in the specimen, and the specimen is not uniformly sheared. Shear strain is non-uniform and cannot be measured by this test.

The shear box only measures strength; it does not provide the stress-strain relationship needed for computing displacements of a mass of soil. The simple shear apparatus is devised to shear specimens and is a research tool. The apparatus comprises a rectangular box that is hinged in which a specimen of soil is sheared in a vertical plane changing from a rectangle to a parallelogram. It is seldom used commercially. By far the most popular test to determine stress-strain relationships for soil specimens is the conventional triaxial test.

Some simple testing devices are commonly used to quickly measure shear strength of small specimens of soil. These include a hand-operated vane tester, which comprise a small vane pushed into a specimen of soil and turned; the applied torque is measured. Often a peak strength is measured first, and a residual strength is measured after several rotations of the vane. Another device is a pocket penetrometer, which comprises a small diameter rod that is pushed a set distance into the surface of a specimen causing an indentation. The force that is required to indent the specimen is measured and calibrated on the basis that the bearing capacity at the surface of soil is five times its undrained shear strength. The punched indentation amounts to a small bearing capacity failure on the surface of the specimen. A third device is the fall cone, which was first devised in 1915. A small polished metal cone weighing 60g is placed to touch the specimen of soil. The cone is released, and it sinks into the surface of the soil. The strength of the soil is determined from the depth to which the cone penetrates the surface of the specimen. This test is a bearing capacity failure of a tiny conical foundation and is calibrated to make use of the depth of the indentation to determine the shear strength. These three quick and simple tests are relatively inexpensive and can be useful for ground comprising weak soils with variable strength by obtaining many measurements on many specimens.

Conducting basic classification tests on soil samples to determine types of soil and their states is fundamental to site investigation. Classification tests can be correlated to strength; inconsistencies can be examined to learn whether a sample has been disturbed or has unusual characteristics. Classification tests are used to measure the sizes of the particles comprising the soil and the percentages of different sizes that are present. Coarse-grained particles are sorted by sieving; fine-grained particles are determined from their different rates of sedimentation from a suspension in water. Water content is measured by weighing a specimen, by drying it in an oven and weighing the dried specimen. Atterberg Limits gauge the plasticity of soil. The two Atterberg Limits are determined by two simple tests. The liquid limit (LL) is determined by remoulding a specimen of soil and adding water if necessary until the soil is a paste and is very weak. The procedure is described in BS EN ISO 17892-12:2018B[24]. A small standard cone, as described earlier for measuring strength, is allowed to drop and penetrate the specimen of soil; the depth of penetration that is achieved in four seconds is recorded. The test is repeated, four times, by adding water to the sample of soil and remoulding it progressively until the depth of penetration exceeds 20mm. The results are interpolated to determine the water content at which the depth of penetration would be 20mm; this water content is the LL. The soil at this stage is very weak with strength of about 1.5kPa, according to John Atkinson[25], and is of a consistency such that upon the addition of any further water the soil becomes runny like a liquid. The other Atterberg Limit is the plastic limit (PL). This is the water content of the soil, sufficiently low that it crumbles when rolled by hand to a diameter of 3mm. At the PL, the strength of the soil is about 150kPa also according to Atkinson[25]. The plasticity index (PI) is the difference between the LL and the PL. The larger the difference, the greater the plasticity of the soil. The tests are applicable to plastic soils that are fine grained, such as silt and clay, or mixtures containing proportions of silt or clay. Friable soils such as sand with little or no content of fines cannot be rolled to determine the PL and cannot be formed as a steep-sided groove in the LL apparatus; they are therefore defined as non-plastic soils.

The Liquidity Index (LI) of a soil is how much natural moisture content (Mc) is between the liquid limit (LL) and the plastic limit (PL), as defined in Equation 2.2.

$$LI = (Mc - PL)/PI \qquad (2.2)$$

By assuming that the strength is 1.5kPa at the LL and 150kPa at the PL, as a rough estimate the LI can be used to estimate the shear strength, c_u, proportionately between these two values, as defined in Equation 2.3.

$$C_u = 1.5 + LI \times (150 - 1.5) \text{kPa} \qquad (2.3)$$

Points to watch out for relating to tests for strength of soils in addition to fundamental quality control are sample disturbance (discussed above); air entrainment, which can be a problem for porous soils; anisotropy (whereby strength measured vertically may differ to strength measured horizontally; retained soil develops large shear stresses and would eventually fail on surfaces that are inclined at about 45 degrees to 65 degrees to the horizontal) and strain rate. Anisotropy is primarily a feature of sedimentary soils. If required, specimens can be trimmed and tested at different orientations. The strengths of several types of soil are affected by rate of strain. Quick tests at fast rates of strain can exhibit higher strengths than tests conducted at slower rates of strain. In soft clays, after a period of time, the strength can be reduced to 80% of the quick strength. This phenomenon was apparent in the 1960's a few embankments carefully designed to a factor of 1.2 remained stable for a few weeks and then collapsed unexpectedly.

2.3.3 Stiffness of soil

Determination of strength of the ground is necessary to ensure that failure of a deep excavation does not occur. Ideally, the strength of soil should not be relied upon; designs should ensure no failure by imposing a factor of safety. However, a more detailed evaluation of the effects of deep excavating indicates that at some locations the full strength of soil may be mobilised but the zone where failure is reached is sufficiently small that gross deformation and overall failure do not occur. As excavation takes place, the stiffness of soil is mobilised; the stiffness of soil determines the deformations of the ground, the interaction with lateral earth support and the induced movements of adjacent ground, nearby utilities and buildings. A fundamental requirement is that during construction the ground should not move enough to cause damage or reduction in the level of service within the surrounding area. For this purpose, the stiffness of the soil has to be evaluated.

Stiffness of soil can be measured either in situ in the ground or on samples in a laboratory. In a laboratory, the usual method to determine stiffness is to record the deformation of specimens of soil when loaded in a triaxial test when also determining the strength of the specimen.

A description of field tests for determining the properties of soil is given by Geoguide 2[14]. Testing equipment includes flat plate dilatometer, cone penetrometer, standard penetration sampler and pressuremeters. The SPT measures, in blow-counts, the resistance to penetration, and the CPT measures the tip resistance to a steady rate of penetration; both are indirect ways to measure stiffness of the ground although they are commonly used. The dilatometer and pressuremeters measure the resistance of the ground to lateral displacement. The flat plate dilatometer measures the lateral expansion of two parallel plates that are pushed apart. The first stage of expansion until the soil yields measures the stiffness of the adjacent soil. Dilatometers that are pressed into soil beneath the base of a borehole are intended to cause little disturbance to soil before the start of the test. Two types of pressuremeters were devised for the same purpose by expanding the cylindrical cavity in the ground. However, expansion of a cylindrical cavity very quickly results in yield due to circumferential expansion as the radius of the cavity is expanded. For this reason, tests using a pressuremeter are often subjected to a loading-unloading-reloading cycle because unloading and reloading are assumed to not cause the soil to yield, and the deflections and changes in pressure are used to compute the stiffness of the soil. The tests should be conducted and interpreted carefully because, if the in-situ pressures in the ground are exceeded then the soil will have yielded, and after yielding the stiffness can be very small indeed (Mair et al. [19]).

2.3.4 Coefficient of subgrade reaction

Many years ago, when engineers started to compute deformations of soil, plate-loading tests, either vertically at the surface or laterally below ground were interpreted as a stiffness, called subgrade reaction. Short-term loading and unloading for small deflections was approximately reversible and the coefficient of subgrade reaction k_v vertically and k_h horizontally was calculated as the pressure on the test area divided by the deflection. A horizontally applied pressure P_h resulting in a deflection d_h, was used to determine k_h as defined in Equation 2.4.

$$k_h = P_h/d_h \qquad (2.4)$$

Values of k_h have often been determined by testing with a small plate, say 150mm or 300mm square, but when the values so derived for the same ground were compared with deflections of much larger full-scale footings, the values of k_h were found to be different and the existence of a scale effect became apparent. In simple terms, a small plate is used to load the ground to a small depth whereas a large footing loads the ground to a larger depth. A change of stress results in strain that when multiplied by the depth over which the strain applies results in a displacement. Under the same applied stress, the deflection of a large footing is more than the deflection of a small plate and therefore the stiffness (i.e. loading required per unit displacement) for a large plate is less than the stiffness for a small plate. When using k_h to determine the subgrade reaction, it is necessary to consider the extent of the loaded area and reduce the value of k_h in accordance with the extent of the loaded area. Otherwise, the deflections in response to loading will be underestimated. This was a problem when the "beam on springs" numerical model, explained in Section 3.1.1, was used at first and high values for k_h (obtained directly from a plate-loading test) were input as spring restraints at nodes at various depths of a wall irrespective of the spacing of the nodes, which were generally much wider apart than the width of the plate that was used for the test. The concept of a coefficient of lateral subgrade reaction is still used in some software, but the software is written to take into account the size effect and other effects.

The assumption that swelling and compression is reversible and that k_h is a constant for all changes in loading is a simplification. For example, for larger displacements as the limiting active pressure is approached, the stiffness reduces and, at the limiting active pressure, the soil yields with no incremental stiffness in the direction of yielding. At such conditions, the value of k_h drops to zero. The same applies when the lateral stress is increased to the passive pressure and the value of k_h reduces to zero. Comparing loading and unloading, the strains are not reversible; nonlinearity increases as the stress level gets closer to active and passive stresses in the soil.

Another factor is drainage. If the conductivity of the ground to water is low, the ground may remain undrained during the time for excavation and until the construction within the excavation is completed. In this case, undrained strength and undrained stiffness may be adopted and reasonably represent the performance of the ground. If the ground is highly conductive, drainage may be completed during time for the excavation and drained parameters may be used to estimate the performance of the ground. For ground with intermediate conductivity, the drainage may be only partial during the time of the excavation in which case the degree of drainage that takes place and the resulting uneven pore water pressures in the ground should be computed; deformations should also be computed using drained parameters. As an alternative, two sets of calculations may be performed assuming completely undrained and completely drained conditions, respectively, and the more onerous result may be used for the design. Commercial computer programs are available whereby partial drainage can be numerically modelled and the displacements of the ground, loads in struts and bending moments and shear forces in retaining walls are computed. In an arbitration where drained, partially drained and fully drained analyses were carried out, it was not brought to light until the last moment that the partially drained case was almost fully drained. However, the difference in loadings on the lateral earth retaining system compared to the drained analysis were not due to incomplete drainage but were due to two different representations of the retaining wall and drainage system. In the one case, the wall was modelled with impermeable wall-type elements; in the other case the wall was modelled as a beam resting on the freely draining face of the excavation. The adopted boundary conditions were not the same.

That said, for many designs, the assumption that changes in lateral stresses can be estimated using the coefficient of lateral subgrade reaction, k_h, can be sufficient or assuming that the soil has bulk moduli M, for isotropic stiffness, and G, for shear stiffness, and behaves as an elastic continuum until the limiting active and passive pressures are reached can be appropriate provided that realistic values are chosen for the parameters.

If the ground is sufficiently permeable for some drainage to take place during the lifetime of a deep excavation, values are needed for parameters for permeability, and consolidation and swelling. The parameters are permeability, k, and coefficients of compression such as C_c for virgin consolidation, C_r for unloading, C_{rc} for re-loading and P' for the pre-consolidation pressure above which primary consolidation takes place and below which swelling and re-consolidation takes place. Coefficients C_c, C_r, C_{cr} all related volumetric strain to the logarithm of applied effective stress. The logarithmic function shows that soil gets less compressible at higher confining pressures. However, the logarithmic function is an approximation that is useful only within the range of engineering applications. Beyond such a range, a negative logarithm trends towards negative infinity, whereas soil cannot be compressed to a negative volume. Consolidation parameters can be determined from drained tests in a triaxial apparatus but are generally determined in oedometer tests.

For simplicity of analysis, sometimes pseudo-elastic parameters are determined from tests. A more straightforward representation of the stiffness of the ground is to determine the bulk moduli for M for volumetric stiffness and G for shear stiffness. It is now a common practice to analyse test results to compute M and G moduli and use the values in more detailed calculations to determine ground movements in response to changing loading.

2.3.5 Soil pressure

During the excavation process the temporary ground retaining structures experience lateral pressures from the ground; they have to be designed and built with sufficient strength to fully support the ground with a factor of safety. It turns out to be convenient to consider pressures in terms of total stress, effective stress and pore water pressure. These terms are straightforward. Total stress, say, immediately beneath a loaded area such as a shallow footing, is the total load on the loaded area divided by the area. Pressure in kilopascals (kPa) is load in kilonewtons (kN) per square metre. At depth below the ground surface with no superimposed loads, the vertical total stress is the total density times the volume of soil above the area concerned divided by the area. The total vertical stress is the total mass density of the soil, γ_t, times the height of the soil above the area of concern, h, times the area divided by the area. Hence, independent of the area, the total vertical stress, σ^t_v at a point is the density of the soil times the depth to the point of concern, as defined in Equation 2.5.

$$\sigma^t_v = \gamma_t \times h \tag{2.5}$$

It is helpful to consider separately the stresses in the ground water and the effective stress on the soil, which is the sum of inter-particle forces divided by the cross-sectional area of concern. Water has no shearing strength, and its volumetric compressibility is very low; by contrast, soil particles that are in contact with each other, sometimes called the soil skeleton, can support loads by inter-particle forces and the soil skeleton can deform in response to changes of loading. A soil skeleton has both shear stiffness and volumetric compressibility. Microscopic analysis shows that a soil skeleton deforms overall by a combination of sliding, and breakage of particles primarily at the points of contact and virtually no deformation of the particles themselves. Whereas water is almost incompressible volumetrically, the soil skeleton can be compressed when it is loaded, and it can swell when loading is removed. Water (hydraulic) pressure, σ_w, is the product of the density of water, which is denoted γ_w, and h_w, which is the height of water above the point of concern as defined in Equation 2.6.

$$\sigma_w = \gamma_w \times h_w \tag{2.6}$$

The vertical effective stress on the soil, σ'_v is the difference between the total vertical stress, σ^t_v, and the water pressure, σ_w as defined in Equation 2.7.

$$\sigma'_v = (\sigma^t_v - \sigma_w) = (\gamma_t \times h) - (\gamma_w \times h_w) \qquad (2.7)$$

The hydrostatic pressure is uniform in every direction. The soil effective stress is usually not equal in every direction. The ratio of the vertical effective stress, σ'_v, to the horizontal effective stress, σ'_h, is called the lateral coefficient of earth pressure; the undisturbed in-situ value, k_o, is called the coefficient of lateral earth pressure at rest. For soils that have been normally consolidated, that is have been consolidated under the present vertical effective stress, the value is about 0.5. Higher values for k_o are measured in over-consolidated soils, that is soils that have been subjected to a previous additional loading that has since been removed, the value of k_o is greater than 0.5 and can be more than 1.0 or 2.0 depending on the previous maximum vertical effective stress to which the soil has been subjected. Empirically it has been found that for normally consolidated soils, the value of k_o is about $(1 - \mathrm{Sin}\ \phi)$ and this value is commonly used.

Soil that has a history of being consolidated to one pressure and then is allowed to swell to a lower pressure (e.g. by removal of overburden) is said to be over consolidated. The degree of over consolidation is the ratio of the previous maximum vertical effective stress to the current effective vertical stress. The degree of over consolidation affects the value for k_o. The degree of over consolidation can be computed by conducting a consolidation test on a specimen of the soil. The apparatus is called an oedometer; it comprises a stiff ring inside which a specimen of soil is tightly fitted. The specimen sits on a porous base and a porous piston is placed on top. The piston is loaded to successive values of vertical stress, and the specimen is allowed to drain at each stage. When the vertical compression is plotted against the logarithm of applied vertical stress, the results appear to lie on two straight lines connected by a curve. Extrapolation of the two straight lines indicates a turning point from the over-consolidated state to the normally consolidated state. The vertical effective stress at this point is considered to have been a previous consolidation pressure, σ'_{vp}, and the ratio of σ'_{vp} to the in-situ vertical effective stress in the field is the over-consolidation ratio. If it has not been measured, the in-situ horizontal effective stress is assumed to be of the order of k_o times the past vertical effective stress σ'_{vp}.

2.3.6 Limiting pressures

The lateral earth pressure on a retaining wall depends on the lateral movement of the wall relative to the ground, as depicted in Figure 2.6. When the wall does not move, pressure is described as "at-rest" pressure, σ_h. When the wall moves away from the soil, the soil pressure acting on

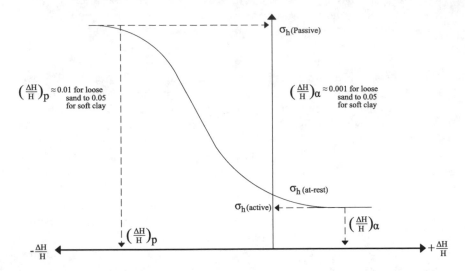

Figure 2.6 Lateral earth pressure mobilisation.

the wall drops to a limiting "active" soil pressure, σ_a. If the wall moves towards the mass of soil, the soil pressure builds up to a limiting "passive" pressure, σ_p.

The great physicist, William J. M. Rankine[7] derived an equation for forces on a retaining wall in the 1800s. When the back of a wall is frictionless and retains dry cohesionless soil with mass density of γ^t and an angle of friction, ϕ and then the wall moves forwards, the soil yields and exerts a horizontal stress σ_a, called the active soil pressure, given by Equation (2.8).

$$\sigma_a = \gamma^t \times h \times (1 - \sin \phi)/(1 + \sin \phi) \qquad (2.8)$$

The vertical stress, σ_v, in this situation is $(\gamma^t \times h)$, and the ratio of horizontal stress to vertical stress, k_a, is called is called the coefficient of active soil pressure. By inspection, k_a is given by Equation (2.9).

$$k_a = (1 - \sin \phi)/(1 + \sin \phi) \qquad (2.9)$$

Rankine also derived equations considering sloping supported ground; when a retaining structure moves in the reverse direction towards the retained soil, like a bulldozer pushing soil, the soil pressure rises to a passive pressure, σ_p, which is greater than the vertical stress, σ_v, and is given by Equation (2.10).

$$\sigma_p = \sigma_v \times (1 + \sin \phi)/(1 - \sin \phi) \qquad (2.10)$$

The coefficient of passive pressure, k_p, is the ratio of passive pressure to vertical pressure; values for k_p are of the order of 3.0 and above. Rankine's methods are generally used in computations of active and passive pressures.

It has been found that when there is a water table in the retained soil, then k_a applies to the effective stresses in the soil. The water pressure remains hydrostatic; it is added to the effective stress to determine the total stress on the wall at each level.

For soils with friction and cohesion, equations that are generally used and attributed to Bell are as reproduced as Equations 2.11 and 2.12.

$$\sigma_a = k_a \times \sigma_v - 2c(k_a)^{1/2} \qquad (2.11)$$

$$\sigma_h = k_p \times \sigma_v + 2c(k_p)^{1/2} \qquad (2.12)$$

In-situ lateral stresses can be measured using the plate dilatometer and the two types of pressuremeter that are described earlier. However, computing the in-situ stresses requires skill. It requires identifying small changes in response to the changes in pressure that are applied to the soil by the testing apparatus. In the case of the plate dilatometer, the soil to either side may be compressed by the insertion of the plate, and the measured horizontal stress may be higher than the in-situ value. The Menard pressuremeter is placed within a borehole where the soil at the perimeter of the bore has been unloaded and the measured lateral stress could be less than the in-situ stress before the borehole was drilled. The self-boring pressuremeter is aimed at inserting the probe with no change in the in-situ horizontal stress in the soil. In practice, in-situ horizontal soil pressures are seldom measured. For purposes of design, values of initial in-situ horizontal stress are usually computed based on the Rankine method for the at-rest pressure as described in this Section making use of the frictional strength of the soils present and the assessed degree of over consolidation.

During deep excavations for the Chicago subway in 1939–42, Ralph Peck[26], as a student to Karl Terzaghi[27], monitored loads in timber struts that were used to brace timber lagging that provided earth lateral support to the excavations. Peck[28] plotted the site data and produced two envelopes of deduced equivalent soil pressure; one envelope was for soil extending below the final excavated level and one for hard ground below the final excavated level. These so-called Terzaghi and Peck envelopes, were used by many designers to compute loads in struts: each strut took the pressure depicted by the relevant envelope between the two levels located midway between the subject strut and the strut above and the strut below. The envelopes were used for computing strut forced for braced excavations for nearly 40 years. I am not aware of any collapse of braced excavations where the Terzaghi and Peck envelopes were used and the loads in the struts were found to have been underestimated sufficiently to cause the collapse. The envelopes, as illustrated in Figure 2.7, are simple to use, and a hand calculation quickly derives loads for each strut in a multi-strutted deep excavation. Such a hand calculation provides a convenient means of checking the approximate magnitude of loads in struts that now-adays are determined by versatile but complicated computer programmes as described in Section 3.1. However, simple hand calculations seem to be not popular now and are a thing of the past.

The Terzaghi and Peck envelopes include data from many struts with dif-ferent spacing and pre-loading; therefore, they can be considered conserva-tive for any particular configuration of struts. By conducting a bespoke analysis for a particular location, the maximum loads in the struts are gener-ally within the envelope and the specific design can be adopted with the pos-sibility of some saving. However, the loading is derived from excavation in dry soil with lagging. The lagging is not continuous; it is erected piecemeal as the excavation is taken deeper. The envelope is not intended to be used as if it were an applied loading to walls that are continuous for the depth of a deep excavation. For deep walls with several levels of strutting, each strut

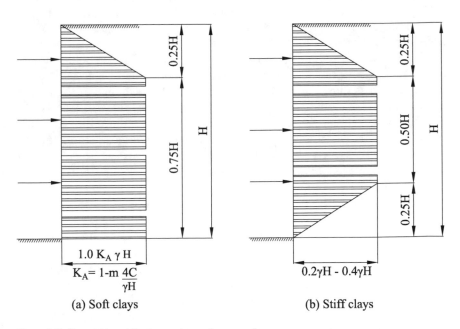

Figure 2.7 Terzaghi and Peck envelopes for strut forces.

is subjected to loading, which is at a maximum when excavation has extended below the strut to the level of the next lower strut before the next lower strut is placed and loaded. When the next lowest strut is loaded, the loading in the strut immediately above is partially relieved due to the load in the lower strut. At all stages, the higher-level struts have been partially released from their maximum loading and the loading distribution on the walls is therefore lower than a combination of the maximum load considered to act on all of the struts that are in place. At first sight, it would appear that the analyses of deep retaining walls (considering the applications of higher combinations of loads than are expected to be applied during construction) would be conservative. This is the case if all of the stages of excavation are analysed and if at each stage the deflections from the previous stage are carried forward. A series of one-off calculations for each stage assuming the maximum loads derived from the Terzaghi and Peck envelope on an initially un-deformed and therefore initially unloaded wall structural element is unconservative; during multi-stage excavations, the lower parts of retaining walls move progressively forwards stage by stage. Sagging bending moments due to deflections towards the open excavation as would apply between struts are greater than when walls are assumed to be un-deformed at the beginning of each stage. Hogging bending moments, in the reverse direction, such as could be developed across the back of a strut, are much smaller when an analysis considers the built-in deformation from previous stages of excavation.

2.3.7 Development of lateral soil pressure

To sum up, before excavation, lateral pressures in the ground at rest are water pressure plus k_o times the vertical effective stress. If as the result of excavation a retaining wall moves forwards, the lateral pressures on the outside of the wall will drop until a limiting active pressure of water pressure plus k_a times the vertical effective stress is reached. In simple analyses a coefficient of horizontal subgrade reaction is used to determine changes in lateral stress resulting from the estimated lateral deflections as a wall moves forward. If the direction of movement of a wall is for some reason reversed, the lateral pressure increases with an upper limit of the passive pressure (i.e. water pressure plus k_p times the vertical effective stress). For deep walls embedded below the excavated level, as excavation takes place, the embedded part of the wall tends to move forwards and the ground in front of the embedded part of the wall below the excavation level resists the movement with passive pressure.

A mistake was made in the past when computers were first used for computations. Some of the first computer programs were written assuming that, at the start of the computation, the soil in front of the embedded part of the wall was at rest with an in-situ lateral stress of $(k_o \times \gamma \times h_e)$

when h_e is the depth below excavation level. This was erroneous because before excavation, the in-situ lateral stress was $(k_o \times \gamma \times (h_e + H_d))$ where $(H_d + h_e)$ is the depth of excavation below original ground level. The excavation process unloads the ground at and below the excavated level. Oedometer tests with gauges to measure lateral stresses have been used for consolidation followed by unloading[29]. It was observed that at the maximum vertical consolidation stress, σ_{vmax}, the horizontal stress, σ_{hmax}, was $(k_o \times \sigma_{v\,max})$. As the vertical stress, was reduced the lateral stress remained almost unaltered until the lateral stress was nearly equal to the passive lateral earth pressure under the reduced vertical stress, σ_{vc}, as described in Equation 2.13.

$$(k_o \times \sigma_{v\,max}) = (k_p \times \sigma_{vc}) \tag{2.13}$$

Manipulating this equation results in an equation to determine the reduced stress that induces passive failure of the soil by one-dimensional vertical swelling, as shown in Equation 2.14.

$$\sigma_{vc} = (k_o \times \sigma_{v\,max})/k_p \tag{2.14}$$

In front of a wall, as the excavation is taken down from a higher level, and before the wall is considered to move, the unloading is considered to be one-dimensional and, due to the one-dimensional unloading, the soil yields and develops the passive pressure down to a depth where the horizontal stress prior to excavation equals the passive pressure developable under the reduced vertical stress resulting from the excavation. Thus, one-dimensional unloading, as in an oedometer, results in developing passive pressure down to a depth that is close to h_p where the vertical stress, σ_{vc}, $(= \gamma \times h_p)$ is $1/k_p$ times the in-situ horizontal stress prior to excavation $(k_o \times \sigma_{v\,max})$ before the displacement of the wall is computed for the current stage of excavation. Then, the wall is considered to move and below h_p the ground resists the forwards movement of the wall; the lateral earth pressure increases in response to the horizontal displacement with an upper limit of the passive pressure, as defined by Equation 2.15. Modern computer programs calculate this effect.

$$h_p = (k_o \times \sigma_{v\,max})/(k_p \times \gamma) \tag{2.15}$$

2.3.8 Numerical modelling

Numerical modelling means assigning parameters to properties of the ground and computing the deformation of the ground when loading is changed. Summarising points made previously, undrained conditions mean that the soil is at a constant volume and each part of the ground is considered to have a cohesive strength, c_u. Constant volume means that the ground is incompressible. Undrained conditions can be represented simply. Calculations adopting total stress, cohesive strength and undrained shear stiffness can be used and volumetric changes are zero. There is a complication when considering elastic parameters. Under conditions of constant volume, Poisson's Ratio, μ, is equal to 0.5 and equations that are based on the theory of elasticity for continua (continuous materials) are indeterminate; as an approximation, in order to manipulate the equations, a value that is very close and just below 0.5 is adopted and the numerical solutions show a very small volumetric change. Drained conditions are more complicated because drainage results in volumetric changes and changes in strength in accordance with changes in effective stress. Computations for changes in volume require values for coefficients of compressibility and permeability. Compressibility determines how much water is expelled from soil when the confining pressure is increased or how much water has to seep into the soil as it swells when the confining pressure is reduced; values for permeability are required to compute the rates of ground water flow. Hence, for a given period of time, rates and quantities of flow of ground water, and commensurate changes in volume of the soil are computed. In the long term, fully drained conditions apply with steady state seepage.

Numerical modelling is a procedure for calculating the performance of an event. For geotechnical engineers, parameters are determined for strength and stiffness; mathematical procedures are used, mostly in computers nowadays, to calculate deformations of the ground consequent upon changes of loading. The roots of numerical modelling of the deformations of the ground are in France, which has produced some eminent mathematicians. The French education system has placed emphasis on mathematics for hundreds of years. As early as 1776 a French military engineer, Coulomb[6], charged with constructing earthworks for defence sought to regularise the design of earthworks. He realised that strength differs for different types of soil and that for stability there should be a factor of safety against exceeding the shear strength of the relevant soil. Coulomb's name has been linked with stability calculation for a long time, and his name remains, linked to Mohr[30], in the Mohr-Coulomb strength envelope for soils. In about 1882, Mohr was the first to relate strength of soils to shear stress.

Numerical modelling of stiffness of materials started when elasticity was measured. Elasticity is the phenomenon whereby material will deform when loaded and return to its initial shape and size when the loading is removed. In 1676, Hooke[31] observed that the extension of a spring is proportional to the magnitude of the applied load and is reversible (Atanackovic et al. [32]). This observation was later dubbed "Hooke's Law." In the 1700's Euler[33] measured elastic properties of solid materials and defined Young's modulus of elasticity, E, which was based on the concept of Hooke's law for linear elements, and adapted the concept to properties of a continuum. E is the constant of proportionality between stress and strain in an elastic medium. It had been observed that when compressed material expands laterally to different extents depending on the properties of the material. The ratio of lateral expansion to axial compression, μ, was given the name "Poisson Ratio" in recognition of a great French Mathematician, Simeon Poisson. There followed a number of closed-form mathematical solutions for the elastic deformation of various solid bodies utilising the two parameters E and μ. In the 1800s, Boussinesq[34] computed stress distribution beneath a footing on the surface of the ground and Timoshenko[35] applied the theory of elasticity for beams as structural elements; his concepts are used for designing earth retaining structures to this day.

However, the stiffness of soil varies and in many places increases with depth. In 1970 Robert Gibson[36] derived some mathematical solutions for elastic material with the Young's Modulus increasing linearly with depth, appropriately named "Gibson Soil." Metals were considered elastic for a wide range of loads. However, metallurgists studied forging (forming metals under very high stresses with permanent deformation called plastic deformation). Prager[37] developed and published mathematical solutions for plastic deformation. One of his solutions was for indentation of a punch into a flat surface. This type of solution is valid for the failure of a footing on penetrating the surface of the ground when the ground fails. Prager's mathematical solutions for plastic deformation of metals are adopted by geotechnical engineers for calculating the ultimate bearing capacity of footings. The same theory is used, for example, in calibrating the pocket penetrometer for measuring shear strength of soils, as described in Section 2.3.2.3, whereby the shear strength is taken to be one fifth of the pressure required to form an indentation on the soil specimen. There have been several other workers on yield criteria such as D.C. Drucker, H. Tresca, R. Von Mises and W. Prandtl. Their contributions are found in two useful textbooks for the interested reader by Prager[37] and by Hill[38].

During my studies for a PhD [39], I developed a method of analysis for a plastic material with strength increasing with depth, a plastic equivalent of Gibson's elastic soil. Such methods of analysis have been superseded by much more powerful finite element and finite difference computer programs as described in Section 3.1.2. There are options for yield criteria in the programs including an option to create one's own. Currently a popular numerical model is strain-hardening, which considers strain-hardening after an initial yield. It also includes strain-softening from initially high stiffness for small strains and a reduction in stiffness prior to yield as is often exhibited by real soils. In commercial computer programs, complicated constitutive relationships are available, and manuals encourage their use and offer default values for parameters. As with any complicated procedure, proper use requires an appropriate level of understanding; would-be users are recommended to attend advanced user's courses in order to understand this type of very useful software.

Combining the theory of elasticity for initial loading and small displacements and the theory of plasticity for large deformations numerical models were developed for elastic-plastic materials. The so-called Mohr Coulomb numerical model of a perfectly elastic material up to a yield criterion, with effective stress parameters c' and ϕ', and then a perfectly plastic material with no incremental shear stiffness, is commonly used to represent soil in numerical models. In the 1950s, Ken Roscoe[40] was challenged by Sir John Baker[41] to design footings that would fail plastically and absorb energy during their deformation to be combined with Sir John's plastically designed steel framed structures. Roscoe was an able mechanical engineer by training who had kept up the spirits of his inmates by teaching university-level mathematics from memory to Prisoners of War prior to release in 1945. Ken Roscoe together with his team including Andrew Schofield, Peter Wroth, R.G. James, John Burland and Edmund Hambly proposed a dilatant soil model for coarse-grained soils called "Granta Gravel" and a comprehensive elasto-plastic numerical model for fine-grained soils called "Cam Clay." With some subsequent refinements, Cam Clay is one of the more refined soil models available commercially today. Together, Granta Gravel and Cam Clay are the basic components of Critical State Soil Mechanics[42].

Engineering properties of soils depend on the degree of drainage. Partially drained conditions arise when there is insufficient time for a mass of soil to drain. Sometimes in the ground there are zones or strata of different types of soil whereby ground with low permeability, during a given period of time, is undrained while permeable ground is drained or partially drained. Software that is available commercially can compute fluid flow (drainage) coupled to effective stress analysis, whereby at given points in time the degree of drainage that would take place can be computed and the corresponding volume changes and shearing displacements are computed. Numerical models can be managed to represent strata of soil with different permeability and therefore have different degrees of drainage and consolidation at successive stages of time.

2.3.9 Factors of safety

Factors of safety come into play when the ground could be loaded to failure. Codes of practice[43] recommend factors of safety for retaining walls. For freestanding retaining walls, there are three modes of failure to be checked. The first is horizontal sliding on the soil beneath the base of the wall. The second is overturning on account of lateral earth pressure on the back of the wall resulting in the wall toppling over the forward edge of the foundation. The third is failure of the bearing capacity of the soil beneath the base of the wall. These simple mechanisms are straight-forward for hand calculations. For deep lateral earth support systems, the retaining walls are usually braced by several layers of strutting; excavation is carried out one stage at a time, and the failure mechanisms are more complicated. For a multi-strutted retaining wall, horizontal sliding is resisted in part by the bracing and in part by the embedded portion of the wall that is below the excavated level. Overturning is also resisted by the bracing. The bearing capacity of the soil below the excavated level is of concern when the weight of the retained soil outside the retaining wall exceeds the bearing capacity of the soil below the bottom of the retaining wall. Failure of soil at this location results in heave of soil within the excavation. Lateral support provided by passive resistance of the soil in front of the embedded portion of the retaining wall is also an issue to be checked. The embedded portion of the wall below the excavated level is pushed forwards by the retained soil that is outside the retaining wall and is resisted by passive pressure of the soil in front of the wall plus the bending moment capacity of the retaining wall at the level of the lowest strut. Failure of this soil and bending the wall results in the bottom portion of the wall kicking in, and the passive zone of soil is heaved upwards. For deep excavations extending below the ground water level, there are two more mechanisms of failure. One mechanism is general uplift when the water pressure at any level below the excavation exceeds the weight of soil that is above the level under consideration. Positive water pressure results in an upwards hydraulic gradient, which reduces the vertical effective stress in the soils above the level. A reduction in vertical effective stress reduces the passive pressure, which is k_p times the vertical effective stress, and a reduced passive pressure can affect the stability of the embedded portion of the retaining wall. The other mechanism is piping failure, which arises when the upwards hydraulic gradient causes the soil below excavation level to liquefy. Close to the embedded portion of the retaining wall, the hydraulic gradient is locally higher than elsewhere. Sometimes the soil is locally more permeable, and piping could occur for that reason. These mechanisms of failure should be checked as appropriate for the successive stages of excavation for every deep excavation project.

There is another means of computing a global factor of safety. This is a facility for some of the computer programs whereby the strength of the soil is progressively lowered until the program ceases to function because of a mathematical instability that normally arises when the computed dis-placements of the model of the soil or deformation of the modelled struc-tures run out of control and a numerical solution cannot be achieved. Continuing deformation amounts to failure of the materials or structure represented by the model. The ratio between the available strength and the reduced strength that results in failure is called the strength-reduction factor of safety. This check usually identifies the mechanism of failure with the least factor of safety. It does not check every mechanism that might be required by a code of practice. Inspection of the number of locations where soil is computed to have reached its shear strength can identify other potential or incipient failure mechanisms.

2.3.10 Soil/structure interaction

Prior to the 1970s, there was little or no consideration of the interaction between the ground and the structures in contact with it for the design of deep excavations. Even the engineering profession was somewhat divided, without interaction. Geotechnical engineers conducted site investigations and measured strengths and stiffness of the ground; they wrote a geotechnical report and handed it over to the structural engineers. Diaphragm walls were not common, and deep excavations in those days were mostly provided with braced steel sheet piling as temporary earth lateral support. Structural engineers used Terzaghi and Peck's envelopes to determine strut forces making use of basic soil strength parameters. Alternatively, structural engineers conservatively used the value for the friction of soil and computed coefficients of earth pressure at-rest, k_o, or adopted $k_o = 0.5$ for fine-grained soil and computed soil pressures and applied factors of safety. Movements of the ground were not calculated on a case-by-case basis. Ralph Peck not only measured strut forces, he also monitored and recorded ground movements alongside deep excavations and plotted them for different distances from excavations of different depths and classified the data into small movements for stiff soils, medium movements for medium strength soils and large movements for weak soils[28]. These charts provided helpful guidance to engineers as to whether ground movements outside an excavation were likely to cause problems to utilities and adjacent properties.

The importance of determining mass stiffness of the ground and using it to predict ground movements was recognised at about 1970. Pioneers included Arthur Marsland[44] who tested large plates set at the bottom of shafts in order to measure stiffness of the ground in bearing and John Burland[45] who used a large water tank founded on chalk at Mumford, a potential site for a CERN Proton Accelerator at the time.

In the 1970s it was recognised that if the face of an excavation moves forwards the lateral pressure drops from the at-rest state to the active state and that the drop of pressure was developed as the wall moves forwards by about 1% to 2% of its height. Model tests conducted by Bransby and Milligan[46] demonstrated the effect. It was also recognised that as a wall moves towards soil, the soil changes from the at-rest condition to the passive state but requires a much larger movement of the order of 5% to 10% of the height of the wall to mobilise the full passive earth pressure (James and Bransby[47]).

As excavations became deeper and more popular, engineers sought better methods of design and contractors wanted to offer competitive tenders, conservative designs of lateral earth retaining structures were found to be wasteful and more competitive designs were sought. For example, if the active pressure is two-thirds of the at-rest pressure there is a potential savings if the mobilised earth pressure (being less than the at-rest pressure) can be determined reliably and used for the design. Therefore, it was necessary to evaluate how much the lateral earth support would move during excavation and how much the soil pressure on the outside of the wall would drop. How much retaining walls move depends on the externally applied soil pressure. Therefore, there was an interaction between the pressure applied by the soil on a structural retaining wall and the deflection of the wall. Likewise, for deep excavations passive soil pressure that develops in front of the embedded portion of the retaining walls develops as the walls move forwards and the support that is required for the embedded portion of wall reduces as the wall bends forwards because it is restrained by the bracing at a higher level and its own structural stiffness in bending. It was also recognised that a deterministic evaluation of deformation of the retaining walls and ground movements would provide a more reliable estimate than would have been the case when using charts of ground movements outside the site and their effects on nearby utilities and property.

The assumption made when considering soil/structure interaction for deep excavations is that a wall that has been constructed from the surface and is embedded in the ground experiences in-situ pressure from the ground on both sides. As soil is excavated on the inside face of the wall, the earth pressure down to the level of the excavation is taken away but the wall remains loaded by the earth pressure on the outside. The wall takes up the difference in pressures as a load and is deflected towards the excavation. As the wall moves towards the excavation, the retained ground on the outside relaxes and the lateral earth pressure drops from the at-rest pressure. The retaining wall comes to an equilibrium supporting a lateral earth pressure from the ground that is less than the in-situ pressure. Conversely, if for some reason, the retaining wall moves outwards then the soil pressure increases until the retaining wall and the soil find a balance. Note that when a strut is installed and pre-loaded and the excavation is advanced to the next deeper level, the retaining wall tends to move forwards below the level of the strut and rotates at the level of the strut such that above strut it can rotate outwards pressing against the retained soil, thereby increasing the earth pressure at that level.

Soil/structure interaction occurs between retaining walls and lateral earth pressure and applies to any mode of deformation of any structure that is in contact with the ground. For example, temporary columns for top-down construction are installed before excavation and before any of the permanent structure resting on it is constructed. Temporary columns that are founded on soil at the base of the excavation interact with the ground. As the structural loading is applied, columns tend to sink. As the site is excavated, a lot of soil is removed and the ground beneath the excavated surface tends to rise. For example, in 1976, temporary columns erected inside hand-dug caissons were adopted for the construction of the Diamond Hill underground railway station, Hong Kong. The footings were 5m deep piers built by hand excavation as an extension of the 1.5m diameter hand-dug caissons; they were founded in CDG. A static load test applying the 400 tonnes working load resulted in a settlement of 105mm. Not surprisingly, concern was expressed by the Independent Checking Engineers that the foundations were too weak, although they had settled by less than 10% of the diameter, which indicated that they had not failed. The designer calculated that surrounding each column location, some 2,000 tonnes of soil would be excavated unloading the CDG at the formation level of the columns and that the columns would rise and not sink as the result of the excavation. Monitoring during construction included cross sections of the north wall, north column, south column and south wall. At the end of excavating, the two walls had risen by less than 5mm and the temporary columns had not sunk under the expected deadweight of 400 tonnes but had risen by about 28mm putting the roof structure in a hogging bending moment instead of a sagging bending moment and sustaining a load of about 500 tonnes each due to the uplifting of the formation soil reacting on the roof and concourse structures[2].

Choi Hung and Diamond Hill underground railway stations were both designed in 1975 as contractor's tender designs with calculations for soil/structure interaction carried out by slide rule and logarithm tables. Although my home office, G. Maunsell & Partners, operated a computer bureau in U.K. at the time, it was not available in Hong Kong. How should one do calculations by hand to allow for soil/structure interaction for multiple stages of excavation? It was assumed that the reinforced concrete structures, namely external walls, roof and concourse floors, would all behave elastically with short-term concrete stiffness. The first stage of excavation commenced with some of the roof structure in place spanning the two walls and resting on temporary columns. For Choi Hung Station the walls were extended with a substantial capping beam; main beams were cast at the locations of the columns acting as struts and the capping beams acted as waling. At-rest lateral earth pressure was assumed to act on the outside of the walls; the excavation inside the station removed the soil down to the level of the concourse. Below excavation level, the walls were restrained on the inside by the unexcavated soil and passive soil pressure was developed at excavation level. Below the passive zone, the lateral earth pressure was initially set at the at-rest lateral earth pressure. These loads were imposed on the structure, which behaved as a simple propped portal frame and the deflections were calculated by using elastic beam theory. The result was a forward deflection of the walls towards the excavation, which reduced the lateral earth pressure from the at-rest pressure. Based on the stiffness of the ground the reduction was estimated, and the reduced pressure was then applied to the portal frame, which resulted in reduced deflection for which the lateral earth pressure was slightly adjusted upwards. The concourse slab was cast, and the shrinkage was estimated. The excavation was carried out below the concourse slab, and the portal frame was considered to move as the walls took up the shrinkage in the concourse slab. The deflection of the walls was used to estimate further reduction in the lateral earth pressure on the outside of the walls. The soil pressures were reduced to zero down to the level of the excavation, below which passive pressure was assumed down to the level where it matched the lateral pressure from the previous stage. The difference in soil pressures, inside and outside the walls, was then applied to the walls to estimate the deflection and bending moments and shear forces due to the stage of excavation; these were added to the deflections bending moments and shear forces that were calculated for the first stage and were considered to be locked in. The procedure for hand calculation was not that complicated and did not take very long.

Once the contract had been awarded, the computer program DIANA was a "beam on springs" program written in house to carry out the calculations for the wall and adjacent soil, as described in Section 3.1.1. The interaction of walls with the roof and concourse structures were represented in the computer model as bending stiffnesses applied to the wall representing the stiffness of the roof and concourse. The wall was modelled as a structural elastic beam; the soil was modelled as springs at a series of nodes going down the wall in a beam on springs configuration, which was adopted by several other people at the time, or soon after. The soil springs had limits representing the active and passive pressures, and the springs were replaced by forces when the active and passive limits were reached. This computer program, completed in 1976, was one of the first, if not the first such program to be put into use for designing deep multi-strutted walls. After 1976, computers became more widespread and computation of soil/structure interaction became easier. Now engineers expect to use computer programs and not to perform calculations by hand. The development and use of computer programs is described in Section 3.1

References

[1] *Buildings (Construction) Regulations Cap 123B*, HKSAR Government, (2012). www.elegislation.gov.hk/hk/cap123B?xpid=ID_1438402645257_002.

[2] Benjamin, A.L., Endicott, L.J., & Blake, R.J. (1978) The Design and Construction of Some Underground Stations for the Hong Kong Mass Transit Railway System. *The Structural Engineer*, Volume 56A Issue 1, pp. 11–20.

[3] PNAP APP-59 Ban on Hand Dug Caissons, Building Department, HKSAR Government, (2009, August). www.bd.gov.hk/doc/en/resources/codes-and-ref erences/practice-notes-and-circular-letters/pnap/signed/APP059se.pdf

[4] PNAP APP-12 Prestressed Ground Anchors in Building Works, Building Department, HKSAR Government, (2009, August). www.bd.gov.hk/doc/en/ resources/codes-and-references/practice-notes-and-circular-letters/pnap/signed/ APP012se.pdf

[5] Bezzano, M., Smith, S., Yiu, J., & Wiltshire, M. Case Study: Design and Construction Challenges for Admiralty Station Expansion. *Proceedings of the HKIE Geotechnical Division Annual Seminar 2017*. www.hkieged.org/down load/as/as2017.pdf

[6] Coulomb, C.A. (1776) Essai sur une application des regles des maximis et minimis a quelques problems de statique realtifs a l'architecture. *Memoires de l'Acadamie Royale des Sciences*, Paris, Volume 7, pp. 343–382.

[7] Rankine, W.J.M. (1857) On the Stability of Loose Earth. *Philosophical Transactions of the Royal Society of London*, Volume 147, pp. 9–27.

[8] Casagrande, A. (1936) Characteristics of Cohesionless Soils Affecting the Stability of Slopes and Earth Fills. *Journal Boston Society of Civil Engineers*, Volume 23, pp. 13–32.

[9] Taylor, D.W. (1948) *Fundamentals of Soil Mechanics*, John Wiley & Sons, New York.

[10] Terzaghi, K. (1943) *Theoretical Soil Mechanics*, John Wiley & Sons, New York.

[11] Holtz, R.D. & Kovacs, W.D. (2011) *An Introduction to Geotechnical Engineering*, 2nd Edition. Prentice Hall, Englewood Cliffs, NJ. ISBN-13:9780132496346.

[12] Atkinson, J.H. & Bransby, P.L. (1977) *The Mechanics of Soils*, McGraw-Hill, New York.

[13] BS 1377. (2019, July) *Methods of Test for Soil for Civil Engineering Purposes*, 6th Edition, BSI Standards Ltd., London. ISBN 978 0 580 93469 8.

[14] Geoguide 2. (1987) *Guide to Site Investigation*, Geotechnical Control Office, Hong Kong.

[15] Louis Menard: French Engineer. www.menard-group.com/en/history/

[16] Self-boring Pressuremeter: Cambridge Insitu Limited. www.cambridge-insitu. com/products/pressuremeters/introduction-pressuremeters

[17] Wroth, C.P. (1929–1991) Professor Oxford University, Cambridge University, Master Emmanuel College Cambridge. https://en.wikipedia.org/wiki/ Peter_Wroth

[18] John W Hughes Research Student University of Cambridge Department of Engineering, (c1970).

[19] Mair, R.J. & Wood, D.M. (1987) *Pressuremeter Testing Methods and Interpretation*, CIRIA Ground Engineering Report, Butterworths, London, ISBN 0-408-02434-8.

[20] Robertson, P.K. & Campanella, R.G. (1983, November) Interpretation of Cone Penetration Tests – Part I (Sand) and Part II (Clay). *Canadian Geotechnical Journal*, Volume 20 Issue 4, pp. 718–733.

[21] Mayne, P.W. (2007) *Cone Penetration Testing, NCHRP Synthesis 368*. Transport Research Board, Washington, DC. ISBN 978-0-309-09784-0.

[22] Head, K.H. (2006) *Manual of Soil Laboratory Testing*, 3rd edition, Whittles Publishing, Dunbeath, Scotland. ISBN-13: 978-1420044676.

[23] Lade, P.V. (2016) *Triaxial Testing Apparatus*, Wiley-Blackwell, Hoboken, NJ. ISBN 9781119106623.

[24] BS EN ISO 17892-12:2018B. (2018) *Geotechnical Investigation and Testing. Laboratory Testing of Soil. Determination of Liquid and Plastic Limits*, BSI Standards Ltd., London. ISBN 978 0 580 94083 5.

[25] Atkinson, J.H. (2007) *The Mechanics of Soils and Foundations*, 2nd Edition. Taylor and Francis, London.

[26] Peck, R.B. (1912–2008) Professor, University of Illinois. www.tunneltalk.com/Obituary-Ralph-Peck.php

[27] von Terzaghi, K. (1883–1963) The Father of Soil Mechanics. https://en.wikipedia.org/wiki/Karl_von_Terzaghi

[28] Peck, R.B. (1969) Deep Excavations and Tunnelling in Soft Ground. *Proc. of the Seventh International Conf. on Soil Mechanics and Foundation Engineering*, Mexico City, 1969 State of the art vol. pp. 225–290.

[29] Endicott, L.J. & Cheung, C.T. (1991) Temporary Earth Support. *Proc. Seminar on Lateral Ground Support Systems*, H.K.I.E., May 1991, pp. 39–49.

[30] Mohr, C.O. (1835–1918) Professor, Stuttgart and Dresden. https://en.wikipedia.org/wiki/Christian_Otto_Mohr

[31] Hooke, R. (1635–1703) Natural Philosopher, Surveyor of London. https://en.wikipedia.org/wiki/Robert_Hooke

[32] Atanackovic, T.M. & Guran, A. (2000) *Theory of Elasticity for Scientists and Engineers*, Birkhäuser, Boston, MA, p. 85. ISBN 978-0-8176-4072-9.

[33] Euler, L. https://en.wikipedia.org/wiki/Leonhard_Euler

[34] Bousinesq, J. (1885) *Application des Potentiels a l'Etude de l'Equilibre et du Mouvement des Solides Elastiques*, Gautier-Vilard, Paris.

[35] Timoshenko, S.P. (1934) *Theory of Elasticity*, McGraw-Hill, New York.

[36] Gibson, R.E. (1967) Some Results Concerning Displacements and Stresses in Non-homogeneous Elastic Half-space. *Geotechnique*, Volume 17 Issue 1, pp. 58–67.

[37] Prager, W. (1959) *An Introduction to Plasticity*, Addison-Wesley, Reading MA.

[38] Hill, R. (1950) *The Mathematical Theory of Plasticity*, Clarendon Press, Oxford. Published: 1950, Revised 06 August 1998. ISBN: 9780198503675.

[39] Endicott, L.J. (1971) *Centrifugal Testing of Soil Slopes*. Ph D thesis, University of Cambridge, 1971.

[40] Roscoe, K.H. (1914–1970) Reader, University of Cambridge. https://en.wikipedia.org/wiki/Kenneth_H._Roscoe

[41] Baker, J., Sir. Head of Department of engineering, University of Cambridge, (c1970).

[42] Schofield, A. & Wroth, P. (1968) *Critical State Soil Mechanics*, McGraw-Hill, London.

[43] BS 8002. (1994) *Code of Practice for Earth Retaining Structures*, BSI Stadards Ltd., London. ISBN: 0 580 22826 6.

[44] Marsland, A. & Randolph, M.F. (1977) Comparisons of the Results from Pressuremeter Tests and Large in Situ Plate Tests in London Clay. *Geotechnique*, 1977, Volume 27 Issue 2, pp. 217–243 ISSN 0016-8505 | E-ISSN 1751-7656.

[45] Ward, W.H., Burland, J.B., & Gallois, R.W. (1968) Geotechnical Assessment of a Site at Mundford, Norfolk, for a Large Proton Accelerator. *Geotechnique*, Volume 18 Issue 4, pp. 399–431. ISSN 0016-8505 | E-ISSN 1751-7656.

[46] Bransby, P.L. & Milligan, G.W.E. (1975) Soil Deformations near Cantilever Sheet Pile Walls. *Geotechnique*, Volume 25 Issue 2, pp. 175–195.

[47] James, R.G. & Bransby, P.L. (1971) A Velocity Field for Some Passive Earth Pressure Problems. *Geotechnique*, Volume 21 Issue 1, pp. 61–83.

Chapter 3

Design

The design of deep excavations is primarily the design for an Earth Lateral Support (ELS) system. In most cases, the ELS includes a robust wall around the perimeter of the excavation and shoring to limit the deflections of the retaining wall. Other concerns are control of ground water control to limit the drawdown and settlement in the vicinity of the excavation, protection of adjacent buildings from damage and instrumentation to monitor the performance. Sometimes the design of a permanent underground structure, in whole or in part, is included in the design of the ELS. Design of temporary works for construction includes measures for site drainage, stability of temporary cut slopes, and any facilitating works such as temporary decking, site accommodation and diversion of utilities.

3.1 Design process

The design stage is considered by many engineers the most important stage of creating a deep excavation. For civil engineering of deep excavations rather than for open pit mining, a design entails determining what to construct and how much excavation is needed for a construction to take place. However, that is not necessarily an end in itself. Few deep excavations take place in isolation and a design must consider the effects of the works on the surrounding environment including effects on adjacent land and structures as well as not creating a nuisance to the neighbours. The design should also consider the space available for construction and the equipment that is available. The primary structures to be designed in order to be able to excavate are the ELS system. The effects on the surroundings relate to how much the adjacent ground and nearby buildings move and whether the ground water regime is affected by the works. This chapter addresses the methods by which structural design and effects on the surroundings are carried out.

The overall design process includes investigating the ground conditions at the proposed site and determining values for strength and stiffness of the ground as described in the Chapter 2. The next stage is to prepare a preliminary design, if necessary, then a construction method and sequence and then a detailed design. Both preliminary and detailed designs involve structural design for the ELS and estimation of movements of the ELS structures and the surrounding ground at successive stages of the excavation.

Over forty years ago, when I started to design underground structures, before computers came into general use, hand calculations, sometimes assisted by mechanical calculators, were used. For design of earth retaining walls for deep excavations, very often at-rest earth pressures for the ground were adopted as loading on the outside of retaining walls and the Terzaghi and Peck envelope, as described in Section 2.3.6, was a popular method for estimating loads in struts. Deformation of the ground was not assessed and, when it was deemed necessary, ground settlement outside the excavation was estimated, generally by using charts published by Peck [1]. Design calculations for ELS were generally performed by structural engineers. Geotechnical engineers arranged the site investigation and soil testing and provided values for parameters for use in the design. A comprehensive review of fundamental design methods for excavations dating from the times when computers were not commonly in use was published by the Geotechnical Control Office of HKSAR Government (GCO) [2]. In later years, the logic that was used for design has been incorporated into computer software with the addition of more comprehensive constitutive numerical models for the deformation of soils. Pappin et al. [3]. describe methods of design for deep excavations with reference to the use of computers. A guide to retaining wall design was first published by GCO in 1993 and has been continuously updated [4].

3.1.1 First computer programs

In the mid-1970s, the first generation of computer programs was put into use, and these were the "Beam on Springs" type. The programs considered a two-dimensional cross-section of a wall for which a unit width of the wall is represented as a vertical elastic beam. The ground that is in contact with the wall is represented as a series of springs both outside the wall and in front of the embedded portion of the wall below the level of excavation. The contact force at each spring is proportional to the displacement at that point independent of any other loading in the vicinity. This numerical model comprising a number of springs in parallel is generally attributed to Emil Winkler [5] who, in the 1860s, considered the contact pressure of a foundation bearing on the ground. A numerical model of the ground with parallel springs and is called "Winkler Foundation." A description of the Winkler Foundation is given by Kurrer [6]. In the numerical model, the springs are independent of each other and can be loaded to represent the initial lateral earth pressure on either side of the wall. In the computer programs, the stiffness of springs is calculated from the coefficient of horizontal subgrade reaction of the ground, and the elastic range of springs is limited to the range between the active earth pressure and the passive earth pressure. When the active pressure or the passive pressure is mobilised at any point, the springs are replaced by forces representing the limiting earth pressures. As excavation is taken deeper, corresponding springs on the inside face of a wall, as well as earth pressures down to the level of the excavated surface, are taken away and the installed props are modelled as springs with the stiffness of the props.

The purpose of this type of program was to model the sequence of deflections of a retaining wall during successive stages of excavation and to compute bending moments and shear forces in the wall for purposes of structural design at each stage. The first programs of this type took no account of ground water, and the stresses in the ground were computed as total stresses. Ground water conditions were determined by hand calculations, and the input to the program was adjusted accordingly. Later the programs were improved, and effective stresses were used in conjunction with ground water tables, which could be different on either side of the wall thereby accounting for lowering of the water level inside the excavation as the excavation was taken deeper. The programs did not compute ground movements outside the retaining wall, and at that time these were often still determined by hand calculations making use of Peck's charts. Some of the first programs mistakenly assumed that at each stage of excavation the earth pressure below the excavated level would commence at the at-rest pressure making use of k_o times the vertical stress, but that was later corrected.

Many programs were written at the time such as DWall, Parois, WALLAP and Retwall. Of these programs WALLAP [7], with many refinements, is still in use. Maunsell's program DIANA was developed in 1976 by my colleague Roger Wright and used in-house for many years, but the program was not ever marketed.

3.1.2 Continuum models

Almost as soon as beam on spring programs were put into use, more powerful computer programs were being written. The beam on springs model was an approximation useful only up to a point because the springs acted independently, whereas the ground is not a series of individual springs. The ground is a mass, and deflection at any point, say on the face of a wall in contact with the ground, affects the surrounding ground. The next generation of computer programs considers a mass of soil as a continuum. Two approaches were adopted. The Finite Element Method considered the ground as made up from many small elements in touch with each other. The Finite Difference Method assigned nodes at an array of points in the ground and computed the displacement of the nodes. The numerical method involved computing displacement of successive nodes in response to loads applied both externally and internally between nodes. At first, the ground was modelled as elastic material. The stiffness of each element, or between nodes, was calculated and the elements were linked to each other as a mesh. Loads could be applied to the numerical model of the ground, and the computer program calculated how the mesh of elements deformed and the displacements of each element and the stresses on each element were computed. Elastic finite element programs were written initially for structures, Zienkiewicz and Taylor [8]. Soon afterwards Brian Simpson [9] at Cambridge University wrote a program initially considering soil as elastic and later incorporating the concept of effective stress and yield independently. Simpson's work became the basis of Ove Arup's OASYS [10] suite of programs. At about the same time, Peter Cundall at Imperial College London and then in U.S.A. wrote a program treating soil in two dimensions as a continuum with nodes instead of elements defined. Under applied loads, the displacements of the nodes were computed incrementally adjusting the difference until equilibrium was approached at each node. The commercial version of the program, FLAC, was marketed by ITASCA [11]. In 1986, an early version of FLAC was used for the first design for private building development in Hong Kong relying on a computer analysis representing soil as a continuum either by finite differences or by finite elements [12]. The hillside side comprised a cutting with an overall slope of nearly 1:1 with two slopes over 50m high, and there was steep natural terrain above the site. In order to create site access and a platform for construction, two deep single-sided excavations were required, one above the other. The embedding of the upper ELS relied on support from the lower ELS as can be seen in Figure 3.1.

Since those early days, the use of computers to aid engineering design has become widespread. The finite element method has become the more popular of the two methods although both methods are in use. In the AECOM offices in Hong Kong, some 30 licensed copies of PLAXIS, a finite element program, are in use, six of which are the more advanced three-dimensional versions. The first versions of these types of programs were simple, but they have become more detailed as the result of successive updates. Initially water tables were unbalanced, but now programs can model seepage of water for a specified period of time, as well as coupled consolidation resulting from changes in water pressure due to changes of loading and offer a choice of various complicated constitutive models such as Cam Clay and Hardening Soil Model, and the user is able to specify its own constitutive model if required.

Commercial programs have become very versatile and complicated while the input systems have become simplified for ease of use. At the inquiry into the fatal collapse of the Nicoll Highway [13], when it came to light that the program had been used incorrectly for modelling undrained conditions as described above, the Tribunal asked one of the authors of the program, who had been called as an expert, whether the program had become so simple to use that it could be operated by somebody with insufficient expertise or with

Figure 3.1 Deep excavation on a hillside.

insufficient experience. As I recall, the authors intended the program to be easy to use, but in practice there is little or no control over the users.

About thirty years ago when many Engineers started to use computer programs, the Geotechnical Engineering Office (GEO) of the Hong Kong Government wanted to ensure that programs not only did what was intended of them but that engineers could use the programs properly. The GEO circulated set-piece tests for approval of use of computer programs and the operators. Approval of new programs remains but, without a licencing system for computer users, users are not tested. The un-enlightened should be wary because programs mostly offer default values for parameters. If the user does not enter his or her own value, the default value is used, which runs the risk of the results having nothing to do with the deep excavation that is being modelled. The program produces a result that might look plausible but is not likely to be correct. As described in Section 5.2.7, in preparation for an arbitration in Singapore a designer did not realise that his coupled flow analysis was fully drained; He did not recognise the difference between his coupled flow analysis and his fully drained analysis: one case represented the retaining wall as impermeable, and the other showed it as a beam resting on a fully permeable vertical cut slope.

3.2 Ground water

3.2.1 The need to control ground water

Most deep excavations extend below the ground water table. Control of ground water and its effects are important considerations for deep excavations: for both the safety of the excavation and to prevent damage to the surrounding area and nearby structures. Allowing ground water to drain out of the ground, or from a nearby water course, into a deep excavation runs the risk of flooding the excavation, which could hold up the work until a remedy is found and could damage plant and materials. Inflow of ground water is not usually fast enough to be a threat to life. Deep excavations close to a large body of water such as a harbour or canal can breach the retaining system and cause rapid flooding with the possibility of loss of life. Karst terrain with large cavities dissolved from soluble limestone can conduct excessive ground water into a deep excavation. Unwanted water in an excavation generally softens the exposed soil forming the base of the excavation. Softening the formation level impairs work as machines get bogged down in very soft soil. Close to the bottom of the retaining walls, softening the underlying soils can reduce the passive resistance and reduce the support to the retaining walls with the risk of a collapse.

Seepage of ground water into an excavation can also cause problems to the surrounding area. If the ground is very permeable and seepage through it is replenished, flow into a deep excavation could have little effect on the surrounding area. However, that is not often the case. If the ground water that flows into an excavation is not replenished and the ground water level in the vicinity is drawn down, the ground near the excavation is de-watered and subsides. The subsidence is generally greatest adjacent to the excavation and diminishes with distance. There are conductive geological features such as buried watercourses or sheared zones in rock that can be conductive over considerable distances. Subsidence can damage structures themselves, and in urban areas subsidence can damage buried utilities or their connections to buildings as discussed in this section.

3.2.2 Joints between wall panels

In order to control seepage of ground water into deep excavations ELS systems are usually designed to be impermeable. Diaphragm walls and steel sheet pile walls are made of low-permeable or impermeable material. However, the completed walls have joints, and these have potential for leaking. Connections between diaphragm wall panels and between steel sheet piles are usually intended to be impermeable although imperfections in construction can lead to leakage. At first, many diaphragm walls were built using steel tubes as stop-ends to form curved smooth ends to panels. Sometimes the steel stop-end became stuck, and many of the joints leaked if only a little when the joint was clean. In the late 1970s attempts to make water-tight joints with overlapping horizontal steel reinforcement between adjacent panels and PVC cloth to prevent concrete from the first panels covering the horizontal starter bars. By 1980, the detail was briefly very popular in Taiwan, but the detail was less than satisfactory. Often the steel reinforcement sticking out from a primary panel was distorted during excavation for the secondary panel, and the PVC sheets became tangled around the reinforcement. The result was a very messy badly concreted joint that often leaked profusely. The detail was abandoned and a simpler arrangement with a stepped joint and a PVC water bar as shown in Figure 3.2 is now commonly used.

With this detail, seepage through joints is far less common. Steel sheet piles can also be water tight. They are fabricated with clutches along both sides. As long as the clutches are not damaged and the steel sheet piles remain straight when driven, the clutches can be water tight. Some suppliers provide elongated clutches to improve on the water-tightness. Steel sheet piling can be used to build coffer dams in open water because the clutches can be very effective in keeping the water out.

Secant pile walls have a large overlap between piles that reduces leakage between piles, but contiguous piles only nominally touch each other and leak quite a lot more. When secant piles encounter obstructions in the ground, the piles can be deflected; when they are driven, the piles are often out of alignment or have gaps between them for other reasons. In

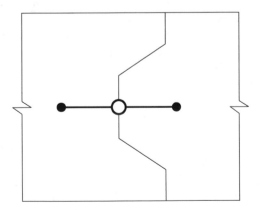

Figure 3.2 Joint for diaphragm walls.

order to make piled walls impermeable, systematic grouting of the ground between piles is normally required. Defective joints in other types of walling may also be sealed by grouting the ground immediately adjoining the leaking joints.

3.2.3 Seepage below walls

It is important to consider seepage of ground water beneath the perimeter retaining walls. When the ground water inside a deep excavation is pumped down to keep the works area dry, if the water level inside the site is lower than it is outside the site and the ground is permeable, water can seep below the retaining walls and up into the excavation and result in problems such as softening or flooding the excavated formation. In order to control such seepage a hydraulic cut-off may be needed, extending below the base of an excavation. In uniformly permeable soils, such a cut-off below excavation level might have to exceed half the depth of the excavation. A hydraulic cut-off can be achieved by extending the retaining walls deeper or by constructing an impermeable or low permeability hydraulic cut-off beneath them. For example, in the case of diaphragm walls, the excavation in slurry-filled trenches might be taken down to the depth required for the hydraulic cut-off and filled to the underside of the diaphragm wall with waterproofing material. Such a hydraulic cut-off can be made with a slurry of cement and bentonite that sets as a mortar or made with plain concrete that can be filled up to the bottom of the reinforced diaphragm wall panel. Above the hydraulic cut-off, the wall panel is then completed with steel reinforcement and structural grade of concrete from that level upwards. Sometimes a hydraulic cut-off is formed by grouting the ground beneath the bottom of a diaphragm wall to reduce the permeability of the ground. The grouting may be carried out below the bottom of the retaining walls via tubes cast in the walls or through grout holes sunk in the ground immediately outside the retaining walls and extending below them to the required bottom of the hydraulic cut-off.

The effectiveness of a hydraulic cut-off must be monitored. Key indicators for protecting against subsidence are drop in water levels outside the site and the onset of subsidence. Small amounts of settlement are almost unnoticed and can be tolerated. When subsidence due to drawdown reaches the limit of acceptability, remedial action may be required. One method of remediation is to carry out more grouting to reduce the conductivity of the ground and reduce the drawdown. However, if the flow of water into the deep excavation is small and can be tolerated, a recharge system can be adopted. Recharging involves having wells between the site and the area that is affected by subsidence and introducing water into the ground via the wells and thereby reducing the drawdown near the wells. This method may be preferred to grouting to save time. Recharging might be preferred or even necessary because the ground water is already flowing because grouting in flowing water can be dispersed by the flowing water and the grouting is then very effective. When drawdown and subsidence are expected to be a problem, recharge wells can be installed before excavation commences. Recharge wells rely on supplying water at a head that is above the pressure of water in the ground. However there is a practical limit in that the head of water in an open well cannot be above ground level or the well will overflow; if the well is sealed and water is supplied from a header tank above ground level, the head of the supply water should not be more than a few metres in order to prevent the escape of water from the wells to the surface thereby creating a quagmire.

3.2.4 Control of ground water seepage

As described above, control for ground water by grouting involves injecting materials into the ground to reduce the permeability of the ground. For many people grouting is pumping cement slurry into the ground. Cement does not dissolve in water. Slurry is a suspension of cement particles in water. Thick slurry is a paste of cement particles with not much water. Using slurry of cement is effective when the pores in the ground between soil particles are bigger than about a half millimetre in order to allow slurry of cement particles to flow through the pores and permeate the ground. Pores of this size occur in clean sand and gravel but not in silt or clay because the silt and clay particles are small and the pores between them are much too small for cement particles to permeate. Cement slurry with a water content of less than one sets hard. The addition of bentonite to cement slurry reduces the strength of the grout when set and makes it more flexible. Cement and bentonite grouts are useful for filling large voids in the ground, but they do not permeate soils with more than about 30% content of silt and clay because of the small size of the pores in the soil. Injection of cement and bentonite slurry creates a bubble or lens of almost pure grout and does not appreciably reduce the permeability of the ground. The addition of sodium silicate to cement and bentonite grout accelerates the setting time but does not materially improve the lack of penetration into most soils.

In order to reduce the permeability of the ground, the grout should fill most of the spaces between soil particles. Water-soluble grout material such as sodium silicate, which has the same viscosity as water, and low viscosity polymers can permeate any soil through which water can permeate. Other chemical grouts such as acrylates can diffuse into fine grained soils, but sodium silicate is used most often. Sodium silicate solution can diffuse into permeable ground; it sets as a gel and reduces the permeability of the ground dramatically. Very fine-grained soils have low permeability, and in order to achieve a lower permeability, injection of grout is relatively slow and the penetration of the grout into the ground before the grout sets is limited. Therefore, injection of grout needs to be at closely spaced points and for sufficient duration so that the soil between the points of injection is permeated and rendered less permeable by the grout. Grouting is controlled by the volume of grouting materials to be injected or by the pressure that is applied to inject the grout. For many applications, the volume of grout is normally the limit and is set to ensure sufficient spreading of grout through the soil. However, if the pressure does not build up, the grout might escape and be wasted. Then, the grouting is stopped, the grout is allowed to set and the point is re-grouted. Grouting pressure is required to inject the grout into the ground. Overall, at the point of injection the pressure applied to the grout should not exceed the overburden pressure (i.e. the total vertical stress), otherwise the grout could escape to the surface, which is not only wasteful but also a nuisance. Escape of grout to the surface often happens and the site area becomes very messy.

When a deep excavation extends below the ground water table, measures for control of ground water during excavation are recommended; it is sometimes very important that the measures are installed before excavation extends below the ground water table. When the excavation extends below the ground water level, if the ground water control system is deficient ground water flows into the excavation and can cause problems. In particular, once ground water starts to flow, grouting in the presence of flowing ground water becomes problematic. Flowing ground water at the point of injection of grout can dilute the grout and make it ineffective, or it can carry the grout away so that it does not set at the required location and does not stem the flow. For almost all applications of grouting to control ground water, it is better to do the grouting when the ground water is static and before it flows. In deep excavations, if a leakage is not discovered until a late stage when the excavation is deep, it sometimes be necessary to flood the excavation in order to stop the inflow and to carry out remedial work. For large excavations, flooding to ground water level could require a lot of water. For example, an underground railway station for an eight-car train has a gross volume of about 60,000 m^3. Sourcing that amount of water and its subsequent disposal would be a major issue. Therefore, it is prudent to conduct pumping tests before excavating to identify any leaks in the retaining walls and rectify them, as described later in this section. In Hong Kong, a full-scale pumping test is a requirement of the approving authority before granting a permit to excavate.

In exceptionally porous ground such as karst and buried rubble mounds, the ground may have to be treated before constructing retaining walls to keep the slurry or water used for sinking retaining walls from escaping and the trench for diaphragm walling or boreholes for piling from collapsing. Where highly permeable ground, such as karst, extends below a proposed excavation, it may be necessary to seal the karst across the base of the excavation to prevent excessive flows of water into the excavation from the highly permeable ground below. Karst is ground, generally limestone, that has been dissolved. Extensive solution leaves behind pillars of rock such as the famous scenery at Gweilin in China. Karst also includes caves and sinkholes, ranging from potholes like the Wookey Hole in Somerset, to the 100m high Batu Caves near Kuala Lumpur. As I recall, the famous Petronas Towers in Kuala Lumpur are located in an area of karst. When the deep excavation for the basement was excavated close to the top of the limestone rock, there was an excessive inflow of ground water that required considerable effort to stem before sealing. Flow of water into deep excavations resulting from karst can lead to rapid development of sinkholes nearby as water flowing through cavity bearing ground can lead to instability of soil spanning over the cavity and soil collapses into the cavity. Recently, I am told, a site investigation borehole lost its flushing water into a suspected cavity in karst and a sinkhole developed alongside the borehole.

3.2.5 Keeping the site dry

For purposes of the stability of the retaining walls and for efficiency, the excavation should be kept dry or well drained if possible. Generally, in a deep excavation, the seepage of ground water and incident rainfall result in free water at the formation level, which should be channelled to sump pits and pumped out of the excavation. On some projects, it has proved effective to install a series of shallow wells below the final formation level and to keep the formation dry. Excavating, transporting and disposing dry soil are so much easier and quicker than for wet soil, especially if the wet soil is sticky.

In many cases where impermeable strata of ground extend to great depths, a complete hydraulic cut-off may not be practical; deep wells can be installed inside the excavation in order to pump water from below the site. Pumping from deep wells can reduce the water pressure beneath the site and thereby ensure the stability of the base of the excavation against uplift or piping failure. Keeping the water pressure low beneath the site can prevent softening of the ground below the excavated level; it also assists in keeping the formation level drained. Deep wells for dewatering can be installed before commencement of excavation and can be used for a pumping test to identify any leakage in the retaining walls. A pumping test should have sufficient wells and pumping capacity to lower the water table within the excavation substantially. Piezometers should be installed between wells to demonstrate the depth of drawdown within the site. A direct method of locating leaks is to install piezometers or standpipes around the outside and close to the retaining wall to monitor the ground water pressure or water table respectively. A drop in water pressure in a piezometer reading indicates a leakage through the wall nearby. Another check is the rate of pumping water in order to achieve the measured drawdown compared to the estimated rate of flow. Excessive flow of water indicates the vicinity where either the wall is leaking or the hydraulic cut-off is imperfect or not deep enough, and remedial work such as more grouting would be required. Very often cost and time concerns limit the number of piezometers or standpipes and then locations of leakages may not be specifically identified because the piezometers or standpipes are too far apart; it then becomes necessary to either install more piezometers or standpipes in that area to better locate the leak or carry out some grouting over the full area defined by the piezometer. Wells used for a pumping test could also be used for dewatering the excavation, provided the wells are deep enough.

A note of caution: some deposits of sand such as are found in large deltaic deposits (e.g. in Kolkata on the massive Ganges delta and at Shanghai on the mighty Yangtze Kiang) can be very deep and are aquifers with massive capacity to yield water. When aquifers are as deep as, say, 100m or more, a hydraulic cut off, say, 30m below formation level for a deep excavation, is insignificant, and pumping from several wells in excavated area is not practical. When the weight of the soil below the formation level and above the aquifer provides a pressure that is less than the water pressure in the aquifer, the base of the excavation would fail by uplift. Digging to such a level or deeper requires special measure to prevent such uplift. Special measures (e.g. digging in small compartments, relying on the surrounding higher area to provide the necessary weight and constructing the base slab in stages including holding down piles) can be effective to limited depths. Another measure is to construct a "buried slab" by overlapping jet grout piles. Diaphragm walls are extended to a specified depth where jet grout slab is cast between the diaphragm walls providing an impermeable layer at sufficient depth that the combined weight of the jet grout slab and the unexcavated soil beneath the formation level is sufficient to provide a factor of safety against uplift from the ground water pressure.

3.3 Protection of nearby buildings and utilities

In densely developed locations, protection of nearby utilities and buildings is an important issue. While excavating, some movement of the ground very close to a deep excavation is inevitable even it is only one or two millimetres. Excessive settlement can damage buildings or utilities. Problems can occur between buildings with different foundations when the buildings subside differentially with respect to each other. Individual buildings can be damaged by differential settlement of the building itself, resulting in tilting, or shearing distortion. Buildings are of many different forms: low rise, high rise; of timber, steel or concrete construction; on shallow footings or piled foundations. Their ability to withstand differential settlement varies. Newly constructed buildings constructed properly are generally robust and can tolerate more movements than some of the older buildings that were constructed with more brittle materials or with materials that have deteriorated over the ages. As an indication, a building in good condition can normally withstand a differential settlement of 1:1,000. Various guidelines have been proposed. A useful study by Burland [14] concluded that a good indicator of the distress of a building is tensile strain in the building. Tensile strain is induced by lateral spread, for example, by the nearby retaining wall moving away from the building; bending when one side develops compressive strain and the other develops tensile strain or by simple shear whereby tensile strain and compressive strain are mutually at right angles and are at 45 degrees to the shearing strain.

There are many types of utilities to consider. Above ground, cables suspended on pylons or poles generally are very flexible. Exposed pipelines on racks, say, near oil refineries, can be readily inspected and monitored. Often the pipes for oil products are of welded steel construction and are fairly flexible, but the consequence of a rupture would be a major fire hazard. Below ground in urban areas there are many utilities, such as pipelines and culverts for storm water drainage; pipelines for sewage, gas and water supply and cables for electricity and communications. The utilities can be of different types of construction such as reinforced concrete, masonry, brickwork, steel, wrought iron, cast iron and PVC plastic. The condition can be very poor due to age and neglect or good if recently laid. Drains and sewers can have perforations where successive connections are not properly sealed. The utilities are often congested, especially in densely developed locations, and the soil around them can be disturbed or loosely backfilled. Some of the lager drainage pipes and culverts may be surrounded by very loose fill or rest on foundations comprising rock rubble, which is highly permeable. Sometimes the flow of ground water outside a culvert through voided ground can be almost as much as the flow inside the culvert. In old areas, many of the utilities can leak and the ground water be contaminated by sewage or industrial effluent resulting in a cocktail of dirty water, hydrogen sulphide, methane gases and other pollutants that can be poisonous or flammable.

For purposes of the design of a deep excavation, a survey of the condition of nearby buildings should be carried out to determine the form of construction of all the buildings that might be affected by the construction work, including the foundations, so that allowable movements for each building can be determined. These limits are then taken into account in the design process to ensure that the ELS in particular, and the control of ground water, result in expected movements that are within the limits. At the start of construction, a detailed building condition survey is carried out to establish and record the condition of the building for reference in case there are any claims for damage or repair after the excavation has been completed. At the start of the design process, a preliminary estimate of the ground movements is prepared and an "influence zone" around the site is determined based on a settlement criterion such as a distortion of

1:1,000. All the buildings within this zone are surveyed and their condition and ability to withstand movements and distortion are estimated. Having estimated the movements that the buildings can withstand, the design is carried out including predictions of movements of the ground and the buildings. For most buildings, the estimates for movements of the ground are carried out for a "Greenfield" site, that is the interaction with the buildings is adopted without modelling. This is generally on the conservative side because the stiffness of a building would reduce the ground movements. A specific analysis is carried out, including modelling interaction with the structure, for only special structures.

The design must be robust enough to control movements to within the limits that the buildings are estimated to be able withstand; otherwise, the design should include remedial or protective work such as propping or under-pinning the susceptible building. In extreme cases, properties in poor condition that are liable to become dangerous may be considered for compulsory purchase. Asking occupants to vacate an occupied building should not be a regular practice. On the other hand, historical buildings such as the Treasury Building, Ritz Hotel, RAC Building and iconic landmarks such as Big Ben in Westminster [15] have to be maintained irrespective of the cost and time required to do so. In Hong Kong conditions were imposed on the development of Pacific Place and "1881." Some mature banyan trees were located within an area that was to be deeply excavated; the trees had to be protected. Structural coffers were constructed around the trees, and the trees were underpinned, which allowed the projects to go ahead. Little do people realise that in the chic shopping mall 1881 they are walking underneath a suspended banyan tree.

The protection of utilities, especially buried utilities, is important. Gas mains could be a fire hazard, and the rupture of pressurised water mains could cause serious flooding. Records of nearby utilities can generally be obtained from the owners of the utilities. Records of sizes, nominal depths and approximate locations are generally available for recent utilities, but records can be imprecise or even non-existent for old utilities. Usually the utility owners have pre-determined limits for allowable settlement and distortion.

Causative factors for ground movement are the lateral deflection of the ELS system and dewatering of the retained ground by seepage into the excavation. Lateral deflection of the ELS induces subsidence and lateral movements of the retained ground. Estimates of subsidence can be obtained from charts published by Peck [1] as described in Section 2.3.6, but there are no similar charts for the lateral movements. Nowadays finite difference or finite element programs are generally used to compute lateral and vertical displacements of the ground at the surface and at depth throughout the model and, in particular, within the ground that is retained outside the retaining walls. These displacements can be used to compare with the allowable limits for the surveyed buildings. If the ground movements extend beyond the area of surveyed buildings, the area can be extended to include further buildings within the zone of influence; these buildings can also be surveyed and assessed.

Drawdown of ground water results in consolidation of compressible soils and subsidence. Generally, drawdown results in lateral movements that are not very large. Drawdown in coarse-grained soils such as sand and gravel results in small subsidence because the soils have small compressibility. Fine-grained soils, especially highly plastic soft clay soils consolidate significantly when subjected to drawdown. The higher the plasticity, the more compressible the soil is. However, such soils tend to have low permeability so the consolidation can be slow to take place depending on the thickness of the consolidating soil. For example, a one metre thick layer of very soft marine clay when subjected to a drawdown of ground water pressure by ten metres at its base would achieve about 90% consolidation in 12 months and settle by about 300mm. Whereas a 2m thick layer of very soft marine clay would take four years to consolidate by 90% and settle by about 600mm. When planning deep excavations in new reclamation, engineers should take account of any regional settlement where primary consolidation is still going on. Drawdown of ground water can be very tricky due to the fickleness of the conductivity of the ground. The amount of drawdown at a given location outside a deep excavation depends on the rate of seepage into the excavation, the duration of the seepage during excavation, the conductivity of the ground and the sources of recharge. If the rate of seepage is small, the ground is relatively uniform and sources of recharge are close by, the drawdown tends to be small and close to the excavation. However, the ground is not uniform. Strata or lenses of permeable ground can act as aquifers; they can transmit water and lower ground water pressures over quite long distances, and layers of low permeability soils can act as aquicludes; aquifers trapped between aquicludes have no immediate means of recharge through the aquicludes, and the drawdown can be transmitted a long distance. As I recall, confined aquifers of this type were evident in Singapore in the early days of constructing the underground railway when shop houses some 200m away from deep excavations were reported to have settled. The unexpectedly long distance was attributed to highly conductive sandy beds of former rivers that lay on clayey subsoil and had been covered with low permeability fill. Ground water response has been observed at even greater distances. Dewatering from a 90m-deep shaft in Tseung Kwan O, Hong Kong resulted in increased settlement of a sea wall that is about 580m away from the shaft in a south/south-easterly direction. Subsequent tunnelling caused more settlement of the sea wall when a tunnel had been driven across the bay and was more than a kilometre away from the sea wall in a west/ north-westerly direction. In this case, the conductive features were thought to have been intersecting minor faults in bedrock. Drawdown continued to increase as the tunnel remained open, and the effects were detected at even greater distances of up to some two kilometres to the north [16].

Since the effects of drawdown can be difficult to predict, it is necessary to limit the amount of water that seeps into deep excavations and to monitor the drawdown during construction. In 1978, during construction of the first line for the underground railway in Hong Kong, several buildings in the Mong Kok District were damaged because they settled to different extents to each other, because they were on different types of foundations [17]. Consequently, the Building Authority (BA) issued a Practice Note [18] requiring designs for ground water control for deep excavations. Since then, a rule of thumb has evolved that the drawdown outside construction sites should not exceed 2m. Since then, reports of damage to buildings due to dewatering in Hong Kong have been very few indeed. Also, in 1978 during the construction of diaphragm walls for the Central underground railway station, the adjacent Supreme Court Building suffered enough distress that it was vacated and later repaired [19], as I recall at great expense. The Supreme Court Building is a landmark colonial building constructed from granite masonry. The masonry is brittle because the large external blocks of granite accommodate lateral movement only by separation at the mortared joints. The building now houses the seat of Government, the LEGCO Chamber. There was some settlement of the building close to the diaphragm walls, but the lateral movement opened several joints widely in the masonry with a total lateral movement stretching the building by about 100mm, as I recall. Subsequently designers were required by the BA to include an allowance for lateral ground movement due to excavation for diaphragm walling. However, I am not aware of any similar occurrence of so much lateral movement during diaphragm walling since then in Hong Kong.

A useful reference on the response of buildings to ground movements induced by excavation can be found in the CIRIA publication SPU SP 199 [20].

Because ground is naturally variable, estimates for ground movement may not be very precise, and very often the conditions of buildings and their ability to withstand movement cannot be accurately ascertained; ultimately the protection of buildings and utilities monitored during construction as discussed in the next section.

3.4 Monitoring during construction

Even with careful investigation and planning, engineering in the ground cannot be fully anticipated. Ground conditions or performance during construction is at best estimated, and monitoring must validate assumptions that were made in design and make necessary adjustments for ground conditions that are encountered. It is common practice for large deep excavations to have hundreds of instruments for monitoring the performance of the work.

For example, West Kowloon Station in Hong Kong was constructed in an excavation that was 30 metres deep, over 200m wide and about 550m long [21]. Instrumentation included the following:

187	Tunnel settlement markers
72	Building settlement markers
90	Inclinometers (in diaphragm walls)
105	Ground settlement markers
19	Inclinometers (in the ground)
11	Piezometers
17	Standpipes
4	Extensometers
15	Tilt markers
22	Vibration monitors

Thirty-five geotechnical engineers were assigned to the Resident Site Staff on behalf of the employer during construction, as I recall.

For deep tunnel excavations, it is common to establish instrumented cross-sections at, say, every 50m. Each cross-section includes two inclinometers, one in each wall; load cells or strain gauges in each strut; settlement survey markers on the top of the walls and on top of any temporary stanchions and surface settlement markers and standpipes at, say, 5m, 10m and 20m from each wall. In addition, any structure that is within the influence zones is usually instrumented. Three settlement markers are the minimum to record settlement and tilt for each building, but in order to monitor distortion at least six settlement markers will be required. Tilt meters are also used on buildings. Some utilities are monitored with settlement markers. Buried utilities may be exposed and have markers attached, although buried utilities are often monitored by surface settlement markers.

Many instruments are monitored automatically and frequently. Monitored data is captured electronically in a data management system and is often available in real time via the internet. Usually the data can be interrogated by the reader who can call up monitoring for given types, locations and periods of time and ask for plots of data versus time or compare one set with another (e.g. ground water levels versus settlement of nearby instruments).

In 1988, many cities were embarking on constructing underground railway systems, and instrumentation was becoming increasingly extensive. Notwithstanding the amount of instrumentation, things did not quite go right on some sites and, when work had to be stopped, there appeared to be no margin of capacity left in the system to allow any remedial works except those that were very costly and time-consuming. Jack Crooks [22] and I were reviewing instrumentation for deep excavations for the Taipei

Metro. The question arose as to how to convey to site supervisory staff the designer's expectation for each instrument and how to know when it was safe to carry on and when to take action. We decided that something like traffic lights with levels of "Stop," "Caution" and "Go ahead" were needed. We decided that each instrument should have limits set to advise site staff. For example, in the case of struts supporting retaining walls, each strut had an allowable capacity determined by its material strength and its slenderness. The load in any strut should not be allowed to exceed its capacity. When a strut was observed to be loaded fully to its capacity and if activities on site would further increase the load in the strut beyond the allowable load, some preventative action or remedy was necessary. Leaving action until the capacity was exceeded left very little room to manoeuvre and probably the work would have to stop. Some earlier alarm should be raised when action could be taken without necessarily stopping the work. We adopted three limits as follows:

i. Check the design assumptions and the trend of increases in the readings and decide if action should be taken to ensure that work can be continued safely. Otherwise, continue to work keeping the monitoring results under review.
ii. Check the design assumptions with the designer and decide what action to take.
iii. Stop work and revise the design or method of working; do any necessary remedial work until it is safe to proceed.

For Taipei Metro, the load limits in a strut are typically of the following order, limit 1 was about 80% of capacity, limit 2 was 90% of capacity and limit 3 was 100% of capacity. No doubt others thought about setting limits at the same time, and three limits have become the norm in the industry and have been named the Alert, Action and Alarm Limits (AAA). The system has been widely adopted and now the second stage, the Action Limit, is often set equal to the estimated performance and the first stage, the Alert Limit, is 80% of the second stage. These stages relate to whether the design assumptions are being met. The third stage, the Alarm Limit, is when the performance goes beyond expectations and some component is going to be loaded in excess of its capacity and work must stop and preventative measures may be necessary. Limits are, or should be, set for every instrument within the site and outside the site on nearby roads and structures. The need to limit the maximum force in a strut to prevent failure of the strut by over-loading it is easily understood. However, a limit on the maximum force primarily concerns the structural capacity of the strut member and does not take into account the designer's intention. The purpose of a strut is to provide support to something such as a retaining wall and is a very important function for the safety of the site and for the safety of the surrounding areas. If a propping force in a strut is too low, it needs to be investigated. If for some reason the forces in the struts are inadequate, the retaining wall does not get enough support and would deflect more than intended or the retaining wall could itself fail. This situation arose when the force measured in the ninth level strut in a deep excavation 27m below the level of the Nicoll Highway in Singapore in April 2004 [23] led to lack of support to the diaphragm wall and eventual collapse. Instrumented struts had strain gauges that automatically monitored the load in the strut every hour. The ninth level strut was preloaded, but after lock-off the force in the strut was only 17% of what it should have been. According to the evidence, this went unnoticed. Excavation continued down to the tenth level and the force in the ninth level strut increased to 35% of the force that was intended by the designer. Reduced capacity in one strut out of nine might not have been a problem, but the forces in the other struts were also less than intended. The total of the measured forces in all nine struts at this location was only two-thirds of the intended amount. One can consider that the factor of safety against lateral displacement for retaining walls is 1.5; therefore, two-thirds of the required restraining forces amounts to a factor of safety of 1.0, which means no margin of safety and imminent failure. By midday on 20th April 2004, an inclinometer at this instrumented cross-section showed over 300mm lateral deflection and bending of the wall in excess of the onset of the first yield of the steel reinforcement in the wall, which was a clear sign of distress. Total collapse ensued at 15:30 hrs. A fuller account of the collapse and lessons learned is presented in Section 5.2.3. The lesson from this part of the story is that attention should be paid not solely to the maximum force in a strut to prevent yielding of the strut itself but also on achieving the minimum force in the strut necessary to provide adequate support to fulfil its function and the designer's intention.

Instrumentation and monitoring are not only means of control; they can be used positively when the Observational Method is adopted as described in the next section.

3.5 The Observational Method

The Observational Method is a very pragmatic means of risk reduction inherent in ground engineering on account of uncertainty about the type of ground, its properties and how it will behave during engineering works. It was first publicised by Peck [24]. The Observational Method is particularly useful for deep excavations where the consequence of unanticipated performance could be of great consequence [25].

The method was devised by Ralph Peck and Karl Terzaghi in the 1930s. The principle is to estimate the ground conditions, predict how the ground and associated structures will perform and prepare a design accordingly.

The design is not good for all time. The designer assumes that the performance of the ground and related structures such as the earth lateral support might be different from the estimate and that the performance of the excavation is monitored in a systematic manner by instrumentation during construction. The monitoring results are compared with the assumptions. If the performance differs from the assumptions, the designer evaluates the reason. The next decision is whether the design or the method of construction should be modified to comply with required standards of safety and protection of property or whether the work can continue with further monitoring. If the monitored performance is not as good as expected remedial measures are proposed and implemented to meet the specified requirements of performance. When using the Observational Method, it is prudent to plan on what performance might be different during construction and what remedial works would then be needed in advance of construction; An additional consideration is whether to have equipment and materials on hand in case they are needed at short notice. In layman's terms, build according to Plan A but be ready to implement Plan B. Readiness to adopt Plan B is important if, for reasons of safety, urgent work is required. It also can be prudent to save time that would be necessary to mobilise equipment and materials should they be required: during construction, time is of the essence and delays can be expensive.

I first adopted the Observational Method for the design and construction of two underground railway stations in Hong Kong in 1975 [26]. In those days, deep excavations were very few indeed; there was little knowledge of the mass engineering properties of the ground. It was prudent to be conservative when, literally, breaking new ground. However, my design was prepared for a construction company who wanted to submit a competitive bid. For three underground railway stations, top-down construction was chosen. Designs were prepared using moderately conservative soil parameters. The company bid competitively, and it was awarded contracts for two stations. Deep walls were constructed from the ground surface by the hand excavation method as described in Section 2.1.6. The first stage of excavation was followed by constructing most of the permanent reinforced concrete roof structure, and the second stage of the excavation was completed by constructing most of the reinforced concrete concourse floor. The design for the third and final stage of excavation to the underside of the track slab included a temporary level of props because of the large span from concourse to final excavation level and the relatively high ground pressures at those depths. During excavation and construction, instrumentation was monitored. Inclinometers were used to monitor the lateral deflections of the walls. By the time the excavation had reached the point at which the temporary struts were to be installed, the monitoring showed less bending of the walls than had been estimated in the design. A back-analysis was conducted by progressively adjusting the soil parameters to match the deflected form of the walls, and then these parameters were used to predict what would happen to the walls if the temporary struts were not adopted. The calculations indicated that the walls would be within prescribed limits for design. The temporary props were not adopted with a cost savings. The excavation was then completed more quickly with increased headroom due to the absence of temporary props by using larger equipment. The stations were completed satisfactorily. The changed conditions had less settlement outside the site and more reduction in ground water pressures than anticipated, which reduced loads on the ELS walls.

3.6 Geological risk

3.6.1 Geological risks are high

Unlike manufactured materials that are produced to specifications, tested and quality assured, the ground is a natural environment. It is naturally variable, and at the very start of a project, the ground is an unknown factor. The geological risk for deep excavations and tunnels can be high. The probability of encountering unforeseen ground conditions can be high, and the consequential costs can be very high. Claims for unforeseen ground conditions can equal the value of the contract for deep basements and can be double or more the value of the contract for tunnels. Geological risk is such an important factor that it warrants a whole section of this book, references to other sections and the repeating of some relevant parts. A message should be told three times: introduce the topic, explain it and summarise. I make no apologies for emphasising risk mitigation. In the later stages of my career, I have spent too much time as an working out why things went wrong, and I have spent too long in court. I cannot emphasise enough the importance of mitigating geological risk. Section 5.2 describes several projects that went wrong and should not have.

It is important that geological risk be evaluated at all stages of a project and mitigation measures be considered, evaluated and adopted where appropriate. For building contracts, once the foundations have been completed, the main risks addressed by the contracts are inclement weather and price fluctuations. Inclement weather is usually addressed by adjusting the contract period to allow days for bad weather; price fluctuations are addressed by making use of published indices for the cost of basic goods. By comparison, geological risk is of higher significance and greater consequence than inclement weather and basic price fluctuations.

There are several areas of mitigation of geological risk. The first involves investigating the site to determine the types of ground that are present, their extent and their properties. In general, the more investigation and testing that is carried out, the lower the risk of uncertainty about the ground conditions. The design should be performed according to appropriate standards

that include factors of safety aimed at the risk of the ground having properties inferior to those identified and assumed and the probability of loadings
being exceeded. Other mitigation measures are the amount of geological
information made available to the tenderers and the form of contract and
how the geological risk is allocated between an employer and a contactor.
During construction supervision and monitoring the performance of the
works, watching out for any departures of performance from the assumptions and responding to them are mitigation measures.

For many geotechnical engineering projects, including deep excavation,
little is known at the beginning about the ground conditions. When the
excavating advances gradually, more of the ground is exposed and monitoring reveals the performance. At the start of a project, the site needs to
be investigated, the types of ground beneath the site need to be identified
and the engineering properties for each type of ground affected by the
proposed works have to be quantified. Site-specific ground investigation is
usually based on vertical boreholes with in-situ testing and sampling.
These can be supplemented by profiling between borehole locations as
described in Section 5.1.2. In principle, the more factual information that
is obtained, the lower the uncertainty about the ground conditions. The
results from GI are mostly assessed subjectively. Existing data about the
ground can be used to supplement site-specific data and thereby expand
the understanding about the ground conditions and reduce uncertainty.
Sometimes in well-developed locations, sufficient data is available to evaluate
parameters about the ground statistically. It goes without saying that the
more investigation including sampling and testing that is carried out, the
lower is the risk of encountering unforeseen ground conditions.

Site data gathered for design is also important for planning the works
and is important information for tenderers who are planning and pricing
the works. Several years ago, employers or engineers required tenderers to
find out about the ground conditions for themselves and mistakenly withheld geological data gathered for purposes of design. Sometimes tenderers
were allowed only a couple of hours to view stacks of geotechnical
reports and were restricted to photocopying only a few pages. The British
courts took a dim view of this practice, and clearly making data available
to tenderers makes sense. Mobile phones with cameras now permit copying important documents quickly but copying about two dozen thick volumes including archived data takes a long time. Electronic transfer of
volumes of data is now a common practice for major underground projects, and access to data reduces risk.

Contracts for constructing deep excavations usually address geological
risk. Two decades ago, employers often allowed for a few percentage price
adjustments for steel, cement and fuel oil but placed all the geological risk
(perhaps more than the contract value) on the contractors. The objective
often back-fired when a contractor met adverse ground conditions he had not

priced for, and many arbitrations were held for large sums of money relating to unforeseen ground conditions. Contracts now tend to be clearer on geological risk. Some items are re-measured with the employer carrying the risk for these items; clarification of baselines for geological risk is provided in some contracts making use of Geotechnical Baseline Reports (GBRs).

Measures to mitigate risk during construction include independent supervision; instrumentation and independent monitoring of the performance of the works; inspection of the standard of completion and limits for protections of the works, the surrounding area and property. Monitoring results are often available on-line, and mobile phones can show real time monitoring so that people are alerted automatically when AAA limits are reached.

Occasionally it is possible to quantify geological risk for new projects, but very often geological risk can only be evaluated qualitatively. Quantitative Risk Assessment (QRA) relies on performance data. For example, the oil industry has practised QRA for a long time and has gathered extensive performance data (e.g. the frequency of failure of a given type of valve) so that the probability of failure of many components in a system or production plant can be summed. Approving authorities set low probabilities of occurrence that shall not be exceeded, and insurers evaluate financial risk, thereby determining the value of premia to be paid for insurance. QRA is generally not used for deep basements; however, as performance data becomes available QRA might be adopted.

It is becoming the practice to compile a risk register for major projects. The objective is to record all possible risks, from cradle to grave, from the planning to design, construction, operation and eventual decommissioning of the project. The procedure is to identify and record the risks, determine means of mitigation for each risk, record the mitigation measures that are required to meet standards of risk, ensure that the measures of mitigation are adopted and track the performance of the facility throughout its operations and final decommissioning. Geological risk is usually one section of the Risk Register; for projects involving deep excavations, most of the entries relate to planning, design and construction stages.

3.6.2 Mitigation of geological risk during planning

Managing geological risk begins at the planning stage by gathering as much data as possible about the site including published geological information and historical records about the site. Available information includes geological maps. At some places, only regional maps are available; these indicate the general geological conditions such as type of rock and approximate traces of major faults. At other locations, available geological maps are very detailed. For example, the Hong Kong Geological Survey gathered geological information from many projects; it has produced geological maps to a scale of 1:5,000 and geological memoires to even smaller scales. In locations of intense development, archives of borehole records and ground investigation reports can provide useful information about the ground conditions near the site. In the case of the re-development of a site, there may even be historical boreholes from previous development. In Hong Kong, GEO established a Geotechnical Information Unit [27] that collected geological and geotechnical information for planning deep excavations and other ground engineering works. Their database of archived boreholes now exceeds 400,000. In addition, as-built record plans are archived in Hong Kong by the Buildings Department (BD). Plans include as-built foundation records, many of which give information about the elevation of rock at the location of the building. A word of caution, some of the old records are not accurate. Within a few weeks of my arrival in Hong Kong in 1975, I visited the ground investigation of a boring machine on a bamboo scaffolding over sea water. According to the schedule, that machine had completed many metres of boring that day. I asked the driller to show me the record for the depths of the different soils that had been encountered. He took out his pocket book and wrote them down, including successive Standard Penetration Test (SPT) values, all from memory, with no samples to be seen. "Caveat emptor," buyer beware; there is no guarantee attached to borings dating from those days or for levels of rock depicted on old drawings.

3.6.3 Mitigation of risk by ground investigation

Ground investigation (GI), including site-specific boreholes, is usually carried out for deep excavations. There are no hard-and-fast rules on how much GI is needed for a deep excavation. The amount depends on the complexity of the geology within the site and the other mitigation measures for geological risk. GI is essentially a sampling process. Typically, a borehole diameter is small, and a sample diameter is about 76mm. Spacing of boreholes at the design stage might be 10m for a basement in a city block or more widely spaced, such as to 50m, for an underground railway station, depending on the anticipated ground conditions. The plan area of one borehole is about 0.0044 m^2. At 10m spacing, one borehole is about 0.0044% of the ground, and if samples of soil 0.3m long are taken every metre, the frequency of sampling by volume is 1 in 75,000. This is quite a small percentage. The reliability of the sampling depends, of course, on the degree of uniformity of the ground. For large geological features such as the sediments of the Taipei Basin, it is well known that the alternate lacustrine deposits of clay and sand are extensive across the basin. The extent of the presence and thickness of the respective layers depends on the relative historical dominance of the Tamsui and Keelung Rivers. Within distances such as a kilometre the types of soil and the thickness of the alternating sand and clay layers do not change noticeably.

It is arguable that a thorough ground investigation of an adjacent site could be used for the project at hand.

For locations in which ground conditions change rapidly, there is a higher geological risk than in the relative tranquillity of lake deposits, and a more detailed site-specific ground investigation would normally be required. For example, in topically weathered rocks, the rock head can be extremely variable, as discussed in Section 5.2.6. Variation in top of the rock in diaphragm walls panels at 6m spacing can be of the order of 20m with 90% probability of occurrence. The same data shows that within a 6m long diaphragm wall panel the variation in elevation of first encounter of rock is 6m with 90% probability of occurrence. It follows that boreholes at 6m spacing, one per panel, cannot reliably determine the profile of rock head inside a 6m wide diaphragm wall trench excavation. Since excavating in rock instead of soil can take ten times as long and cost perhaps ten times as much, the percentage of rock to be excavated with diaphragm walling equipment is a geological risk. Likewise, the amount of rock within the ground for a deep excavation has serious consequences in cost and time and possibly the need to mobilise more or different equipment if the amount of rock to be excavated is more than expected. There have been many claims taken to arbitration concerning unforeseen amounts of rock in excavations for diaphragm walls and for bulk excavation. Weathering profiles vary in severity. Closely spaced boreholes and profiling together with evaluation of weathering features such as corestones and weathered joints below top of rock can assist in assessing the weathering conditions.

Site data is very important for planning the work. Site data collected before putting the large projects out to tender can be voluminous. Within a relatively short time for tendering, gathering and interpreting the site data can be difficult. Therefore, providing site data including subsurface ground investigation data to the tenderers in an easily manageable form is important. Building Information Management (BIM) is a process of creating and managing digital information. BIM data files can be readily transferred and interpreted by commercial software. Site data can be used to create a geological model that provides a better understanding of the site including visualisation of the geology and subsurface data. By making factual data available to tenderers in this way, the data can be readily interpreted and used to prepare cost estimates and programmes for the work. Availability of data significantly reduces risk at the early stages of a project. BIM allows data and information to be quickly captured, processed and shared among different disciplines and parties and subsequently can help make decision. BIM is a relatively new concept and is still evolving.

3.6.4 Mitigation of risk during design

The next step in mitigating geological risk is the margin of safety that is included as a basis of design. Codes of practice and regulations specify factors of safety, or partial factors on loads and on material to ensure that failures do not occur. In reality, the objective is to achieve serviceability under normal conditions for 100 years or thereabouts and an extremely low probability of failure under extreme events. Lower factors of safety can sometimes be adopted during construction because the site is not open to the public; it is occupied by people who are, or should be, briefed as to the activities and safety concerns, and the works are, or should be, monitored under technical surveillance and there is, or should be, an AAA system in place to alert responsible and technically competent people if there is a trend that might lead to untoward movements. In summary, factors of safety for temporary stages of construction can sometimes be less than factors of safety for all time.

Following the collapse of the deep excavation alongside the Nicoll Highway in Singapore in 2004 [13], the COI recommended for temporary deep excavations no reduction from the factors of safety that are required for permanent works because of the consequences of the risks involved. This recommendation has been put into practice by the Building Control Authority (BCA) in Singapore. Risk involves hazard and consequence. Hazard relates to the threat or probability of occurrence, and consequence is the impact of the event, such as physical damage or fatalities resulting from the event. This is evident in the *Manual for Slopes* issued by the GEO of HKSAR Government [28], where the risk for slopes are graded according to the geotechnical hazard and the consequence of failure, for example, low consequence within a country park or high risk alongside a school.

For design of deep excavations, parameters for the properties of the ground must be determined by the designer. Very often for a given type of soil a range of values is obtained from tests. In selecting values for design, there is a risk that the chosen value might be unconservative and that less favourable, or even significantly worse, values might pertain on site. Codes of practice and regulations do not go as far as to quantify the method of selection of values for parameters for soil. Designs based on global factors of safety envisaged typical values that imply values that are expected to pertain nearly always, and the recommended factors of safety would allow for low probability worse values. Limit state codes such as Eurocode 7 [29] propose that moderately conservative values for geotechnical parameters should be used in conjunction with material factors. What is moderately conservative is not defined. It remains for the designer to evaluate the values for him- or herself. If enough data is available, then statistical methods could be adopted. Gradually more and more data is being published. However, using this method to estimate the top 1%, ten results out of 1,000, might be a means of determining the 98th percentile above the mean value, and cutting off the bottom ten results would give the statistical 98th percentile below the mean value, but one needs to be cautious about the reasons for weakness. In zones of decomposition of rock there can be up to two orders of magnitude range in strength for fresh rock to highly weathered rock from, say, 300MPa to

3MPa strength and the percentile should be calculated for the specific grade of rock and not necessarily for all grades of decomposition. In the case of computing pressures from the ground on a large retaining wall, there is a possibility that, because of the large surface area, a wall experiences the sum of pressures of the range of soil that is in contact with it, and therefore the average value of the soil properties could be used, whereas adopting the weakest value implies that the weakest soil occurs over the whole wall, which could be unnecessarily conservative. On the other hand, when considering sliding of a wall on a horizontal surface, a thin horizontal layer of soft soil would be the governing case and the average strength of samples taken within a stratum would be unconservative. Risk related to adoption of values for properties of soil can be mitigated by careful and appropriate selection.

3.6.5 Mitigation of risk in conditions of contract

The next stage in a project where risk mitigation can be adopted is when drafting contract documents. In these documents, the allocation of geological risk between an employer and a contractor should be identified. Two decades ago, many contracts were drawn up with a clause in the conditions of contract whereby a contractor should take all the geological risk. Employers expected to have a fixed price or fixed rates for getting the work done and the project completed. Civil engineering contracts were frequently tendered based on competitive pricing. This was not a satisfactory arrangement because in order to win a contract a tenderer needed to offer the lowest price, which meant not only competitive procurement but also including a low budget for geological risk. Projects such as tunnelling and deep excavations have large geological risk and the prospect of large cost over-runs. Many substantial claims for unforeseen ground conditions resulted and costs of arbitration or court hearings considerably added to the costs. There are two methods of addressing geological risk in contracts nowadays. One method is to re-measure geological risk items. The other method is to provide a GBR. For example, rock is generally more expensive and time-consuming to excavate than soil, whether the excavation is in slurry trenches for diaphragm walls or in the bulk in open excavation. Re-measurement of the volume of excavation only means that a contractor bears all the risk related to the amount of rock that might be encountered when otherwise excavating in soil. By adding an extra-over item for excavation in rock, a contractor is paid extra according to the amount of rock that is encountered and the employer bears the risk of encountering rock. There then remains the extra time necessary to excavate the rock. Often no extra time is allowed for excavating in rock and this amounts to risk sharing. The employer foots the bill and a contractor bears the risk of delay and the need to complete within the contract period. Some contacts have provision for sharing the time element of risk as well as the cost. GBRs were instituted in order to define what range of values for baselined parameters would be foreseen by the contract [30]. The objective was to provide a basis for tendering in order to achieve competitive tendered prices and definitions of the range for geotechnical parameters that the successful contractor should expect to encounter and to provide the baselines that an engineer would take into account when evaluating claims for unforeseen ground conditions. The baselines do not foresee the quantities of materials, for example how much very strong rock and how much very weak rock is foreseen by the contract. Baseline values for strength, up to a maximum or down to a minimum, that are foreseen. "Foreseen" is used in a contractual sense and means "that are envisaged in drafting and administering the contract." Foreseen does not imply that such conditions, high or low, will actually be encountered. The meaning is that ground that is encountered for which the values of the parameter lie within the baselines are deemed foreseen and included in the contract sum. Any ground that is encountered with values for the parameter that lie outside the baseline are not foreseen and the baselines are used in the evaluation of claims for unforeseen ground conditions. GBRs have been used for several contracts in Hong Kong and Singapore. When properly drafted, GBRs have been proven useful. In arbitration, some GBRs appear to have been poorly drafted. The selection of values for baselines depends on the allocation of risk. Adopting a wide range for a baseline means that a contractor takes a lot of the risk. A narrow range means that an employer wants a more competitive price and will bear the risk of low probability events. When enough data is available, baselines can be set in accordance with, for example, 95% probability of occurrence. If an employer is averse to risk, the value for a baseline could include the whole range of the available data plus a margin. In this way, an employer can gauge the degree of geological risk and the consequential financial risk.

Most contracts adopting GBRs allow for the evaluation of claims for cost; conditions of the contract do not usually allow for variation of time for completion. With respect to unforeseen ground conditions, a contractor accepts the time-based risk with a need to accelerate or change his method of working to cope with unforeseen ground conditions, and the employer accepts financial risk for the reimbursement of actual costs. There is a sharing of risk, time for a contractor and cost for an employer. The alternative method of reducing a contractor's geological risk by re-measurement means that a contractor is paid according to contracted rates or sums for every item of work that is certified as done. There should be a rate for every type of work, but when unforeseen ground is encountered there might be disputes about the rates. Re-measurement for excavation, whether in diaphragm wall trenches or in bulk for any type of material addresses changed depths for diaphragm wall panels because the dimensions of the excavation are usually finalised. Re-measurement for excavation in ground and extra-over for rock is normally sufficient compensation for work done.

Time is very important. Undue haste runs the risk of cutting corners on site, and the consequence of a collapse is a long delay. The employer may want fast completion. For large projects with deep excavations, costs are high. Often the value of the site and, perhaps, the cost of a lease premium for the proposed development and financing interest charges mount up, but site constraints such as limited working space at ground level and limited points of access constrain the number and size of plant that can be used and limit their rate of achievement. Setting too short a contract period increases the risk of a delay. Means of acceleration can be extremely expensive and, if the contract period is too short it could be physically impossible to complete within the contract period. I have attended several arbitrations with massive claims relating to impossibility to build within the contract period, but the details are confidential. Setting a realistic contract period mitigates the risk of a delay.

3.6.6 Mitigation of risk at time of tendering

Pre-qualification is a procedure whereby construction companies are entitled to bid for the work. The exercise is intended to ensure that the pre-qualified companies have sufficient resources, both technically and financially, to complete the project. Technical competence includes prior experience of successful completion of works of a similar nature. Technical resources include proposing technical staff with the required degree of experience in the various roles of the contractor including project management, project planning and administration of sub-contractors, technical staff, health and safety. Financial resources are required to demonstrate solvency of the company, levels of insurance and availability of funds to finance the project until payment is received for certified work or to finance additional work or emergencies until liabilities are resolved. Frequent employers, such as governments, usually maintain lists of construction companies that are pre-qualified for certain classes of work. For major projects, specific pre-qualification may begin with a well-publicised invitation to construction companies to apply, with the possibility of two or more companies forming a joint venture in order to bid for, and potentially to build, the project.

The purpose of pre-qualification is to avoid risk of awarding the contract to an inexperienced construction company or to one with inadequate resources.

The next stage of a project is when the contract for construction is open for tendering. Just as a designer needs to understand the ground conditions, tenderers need to understand the ground conditions in order to prepare estimates for cost and time and any risk that he should allow for. It is therefore important that the tenderers should have access to as much information about the site as possible. For example, borehole data and geotechnical factual reports are site information. Without such data, the tenderers are in the dark and can only see the current condition of the suite at ground level. In the past, some employers refused to hand over factual data in case it was considered incorrect or not representative, and if a contractor did not carry out its own site investigation and relied instead on the employer's data, he would claim any consequential costs. The law courts in the U.K. take a dim view of employers who withhold factual data. For large projects, it is now a practice to collate site data and prepare a geological model using a numerical database. Handing such a model over to the tenderers saves a lot of time when preparing the tender. It is far quicker to check such a model than it is to compile a new one and is a better practice than allowing the inspection of hard copy reports or limiting the taking of copies, which could be construed as withholding information. The tender period is often quite short, and it is difficult for a tenderer to assess the data that he can obtain. Handing over a Geotechnical Interpretative Report would facilitate the tenderers, but interpretative reports that are prepared for purposes of design are not necessarily good interpretations for planning construction. The differences of interpretation might be subtle, but in addition to any interpretation for purposes of design the contractor should have an interpretation for purposes of construction. Many contracts require a contractor to submit his own interpretative report at an early stage; sometimes one must be presented with the tender.

Making site data available to tenderers is obviously a means of mitigating geological risk. Providing minimal information increases the risk. As a further means of mitigating risk, often the contracts are awarded based on two sets of points, technical points and points for price. When aiming to avoid getting an inadequately resourced technical bid, technical points might be as high as 80% with 20% for price.

3.6.7 Changing practices on sharing risk

One way of managing risk is to share the risk between an employer and a contractor. In the past, some employers have sought to impose all the geological risk on a contractor. However, arbitrations sometimes awarded large sums relating to the consequences of claimed changed ground conditions, even when the intention of employers was otherwise. Even though some arbitrations have made substantial awards for unforeseen ground conditions, some employers are very resistant to the concept of sharing risk. Changing the practices of battle-weary people can be very difficult, but it can happen. For example, in 1995 a contract was let in Hong Kong to drive 23.6km of deep sewer tunnels [16]. In 1996, the deep sewer tunnel contracts went to arbitration for claims relating to unforeseen ground conditions. There was a large settlement out of court in favour of the HKSAR Government, HK$750 million was reported [31]. There was a re-tendering for replacement contracts [32], and the tunnels were not completed until 2001 [33]. Because of this experience, fundamental changes were adopted for subsequent deep tunnel contracts [34]. These, amongst other recommendations, included more ground investigation and the equitable sharing of geological risk. In preparing contracts for further deep sewer tunnels, more ground investigation was carried out, a geological model was prepared for the project and made available to the tenderers. Key cost items for probing the ground ahead of tunnelling and for grouting the ground to limit inflow of ground water was re-measured. The outcome, yet to be published, appears to have been satisfactory to the employer and the contractors. At the same time, drainage tunnels were constructed for the same employer and GBRs were used. For one of the drainage tunnel contracts strong granite rock, mixed ground of soil and rock, and soil were baselined, and the outcome, yet to be published, appears to have been satisfactory.

3.6.8 Risk in contractual arrangements

Some different contractual arrangements are intended to address geological risk. They include lump sum, re-measurement, partnering and alliancing arrangements, amongst others.

Lump sum, with a fixed price, puts all the risk on a contractor, and an employer takes no risk. The problem with this form of contract is that when the costs or time for construction is increased dramatically, changed ground conditions are perceived and, notwithstanding the fixed price nature of the contract, claims are taken up and vigorously pursued at great expense. A lump sum contract entails a high risk for a contractor; one should not be embarked upon unless the ground conditions are very well defined at the time of tendering.

Re-measurement means a contractor is paid in accordance with the physical measurement of work that has been done. An employer takes the financial risk on the items that are measured. For example, if excavation in rock is more expensive than excavation in soil and the volume of rock that is excavated is more than anticipated, an employer pays more than if the excavation was in soil. A contractor takes the risk of being able to complete excavation of more rock than was envisaged within the contract period. For this purpose, a contractor may have to mobilise more or different equipment without any means of recovering the extra costs or the extra time involved. Since excavation in rock usually comes later in the contract, late realisation of a substantial increase in the quantities of rock could be very expensive. A contractor's risk in this respect can be reduced by allowing a re-rating for changes in quantity beyond, say 25%, from the quantities envisaged at the time of tender.

Construction partnering is a commitment between a project's owner, the consulting engineers and/or architects, contractors and other key project stakeholders to create a cooperative project environment with a team committed to understanding one another. It is similar in objectives to Alliancing [35] but generally is limited to the entities that are closely related to the project such as the Employer, the Contractor and the Engineer. The objective is cooperatively to address issues arising during construction on a "win-win" basis and to mitigate disputes and thereby save the costs of the resolution of disputes. The process has been in use in North America since about 1990. As a means of risk reduction, Partnering is aimed at reducing the consequence of risk, including geological risk.

Alliancing is a practice that has developed in Australia and is a sharing of risks amongst all parties with the objective of getting the best deal overall. It is broadly based. For example, it affects the designer in optimising the design and the possible need to redesign later, and it affects the whole of the supply chain affecting basic costs of materials and equipment. Success of alliancing depends on a cooperative attitude amongst all parties with the intention to resolve issues in the best interests of the project. As a means of risk reduction, Alliancing, like Partnering, is aimed at reducing the consequence of geological risk.

3.6.9 Mitigation of risk during construction

In order to reduce the uncertainty about the conditions of the ground, further GI can be undertaken soon after the start of a contract. The GI is usually specifically geared to the uncertain parameters. For example, instead of boreholes with retrieval of samples for testing, probing is often used. If he contractor expects to encounter rock when excavating trenches for diaphragm wall panels, the occurrence of rock is a risk item. In Hong Kong, it is common practice to probe by pre-drilling each panel. Usually the hole is drilled, and the rate of progress is quite fast in soil. Rock is recognised by the reduced rate of drilling. When rock is massive, probing is often stopped on first encountering the rock. In weathered rocks where corestones and weathered seams below rock head could be present, probing is sometimes continued to record the presence and absence of rock material by the rate of advance of the drilling or the drilling could revert to coring. Both methods would be aimed at measuring the percentage of rock encountered.

During construction, it is important that the assumptions adopted for design are verified by inspection of the ground that is actually encountered. If the ground conditions are found to be different from the assumed basis of design, the design should be reviewed, and any necessary changes are made. Based on experience, when changed conditions are encountered, it may be best to resolve the situation as soon as possible so that all parties know what needs to be done and how the changed work will be paid. Early resolution of changed ground conditions can resolve the issue without resulting in claims, and potentially expensive resolution of claims, and therefore mitigates the consequence.

Monitoring the performance of the ground and the structures is the ultimate safety check of the site and mitigation of risk during the currency of a contract. The basic criterion for monitoring is to ensure that the works are performing as predicted by a designer. Section 3.4 describes the origins of AAA limits that should be set to control the work. If the results of monitoring are different from the predictions, the design and method of construction are reviewed and changed, if necessary. If the monitoring results show that the Action Limit is reached, then the work must stop, and remediation is necessary for the safety of the site and the surrounding area. In order not to reach the stop work limit, the earlier limits of Alert and Alarm are set. It is of fundamental importance that the AAA limits are used properly. Section 5.2.3 includes a description of the fatal collapse of the Nicoll Highway in Singapore; one of the leading causes of the collapse was not stopping work when the stop work limit was exceeded; work continued even though collapse was imminent.

3.7 Regulatory control of public safety

Geological Risk as described in Section 3.6 is primarily a commercial risk affecting the employer and the contractor. Stability of deep excavations protects the lives of the workers during construction and provides safety for, and prevents damage to, the surrounding areas. Damage control and public safety are important issues; they are fundamental to design and construction of deep excavations and are addressed in this section.

Different countries have different practices to ensure public safety in relation to engineering works generally. Most of my experience is in Hong Kong and Singapore, where there is control of building works laid down by ordinances and enforced by BA in Hong Kong and BCA in Singapore. Both countries register persons who are engaged in, and are responsible for, building works and exercise control by requiring submissions of plans for approval, giving consent for construction and ultimately giving an Occupation Permit for the finished works. The registered persons are responsible as individuals; they are subject to disciplinary action (including fines, de-registration and even custodial sentences) by the authority. In Singapore, the BCA controls all buildings works, public and private. In Hong Kong, the BA was appointed originally to control private building works, but the control has been extended to provide reference and guidance for public works, the Airport Authority and the Mass Transit Railway Corporation Limited. For ground engineering, the BA takes advice from the GEO of the HKSAR Government.

In other countries there is control over building works to varying degrees, from relying solely on the framework of common law and standards of good practice to the detailed control over the individuals and their practices that is similar to that exercised in Hong Kong.

In this section, the measures of control that are in place in Hong Kong are discussed. The reader is introduced to the widely ranging extent of good practice that should apply to design and construction of deep excavations irrespective of local regulatory procedures. The discussion is in summary only; for further insight and detailed explanation, refer to the sources at end of this chapter.

The basic legislation is the Buildings Ordinance Cap., 123, (The Ordinance) of which the stated objective is

> To provide for the planning, design and construction of buildings and associated works; to make provision for the rendering safe of dangerous buildings and land; to make provision for regular inspections of buildings and the associated repairs to prevent the buildings from becoming unsafe; and to make provision for matters connected therewith [36].

The Ordinance provides for the BA to register Authorized Persons (APs), Registered Structural Engineers (RSEs), Registered Geotechnical Engineers (RGEs), Registered Inspectors (RIs) and Registered Contractors (RCs). The registered persons are individuals who have passed an examination of competence. Companies are not registered. The Ordinance does not seek to control commercial liability in the event of mal-performance of the registered person. In the case of damages in excess of the personal or insured capacity of the registered person, recourse may be made to civil law for recovery of damages such as from the registered person's employer. The Ordinance sets out the appointment, duties and responsibilities of the registered persons, procedures to follow including means for disciplinary action and appeal. The BD provides the office and staff for the BA. The AP is responsible overall for coordinating the works including the liaison with the BD and managing the input from the RSE, RGE and RC respectively. The roles of the RSE, RGE, RI and RC are self-evidently the elements of structural engineering, geotechnical engineering, inspection and construction respectively. Administration of Building Control [37] is elaborated in Practice Notes for APs, RSEs and RG (PNAPs) [38].

Basic technical standards are set out in Buildings (Construction) Regulations [39]. These standards for design and workmanship concern, inter alia, site investigation, materials, loads, to be used, adjoining buildings or land not to be affected, site formation and bulk excavation in the Scheduled Area, and retaining walls. Technical standards are elaborated in PNAPs Part B – "*Application of the Buildings Ordinance and Regulations*," and Part C – "*Advisory*."

Detailed design standards such as the suite of British Standards [40], and Eurocodes [41] are accepted by BD. For example, BS 1377 Part 2 [42] is a reference for testing the engineering properties of soil, and Eurocode 7 is a recognised standard for geotechnical design [29]. Guidance is also given by GEO through Geoguides, GEO Publications and Technical Guidance Notes [43].

The ordinance prescribes procedures that require an owner of the property to appoint an AP to coordinate the work, an RSE to take responsibility for the structural elements, and an RGE to take responsibility for the geotechnical elements of the project. The Building (Administration) Regulations set out procedures [37]. Plans are to be submitted to BD for approval. For Scheduled Areas with unusually difficult ground conditions or special areas for which special attention is required, plans for ground investigation are to be submitted for approval and consent for the ground investigation works. Practice Note APP-57 states the requirements and criteria for an excavation and lateral support plan, which applies to any excavation, for private building, that is more than 2.5m deep and 5m long [44].

Submission for approval for ELS works shall include the following:

- Existing conditions of site and surroundings;
- Foundation details of adjoining structures;
- A schedule of geotechnical design parameters and assumptions;
- Details and sequence of construction;
- Assessment documents on topography, geology, ground water, utilities, water mains, drains, etc.;
- Design calculations for lateral support systems;
- Assessment of effects on adjacent buildings, ground movements and water table;
- Monitoring particulars for adjacent buildings, ground movements and water table;
- Grouting and pumping test proposal, if necessary.

When plans become Approved Plans, application may be made for consent to build in accordance with the Approved Plans. For difficult projects, BD on behalf of BA, may apply additional conditions to the consent.

The Scheduled Areas are considered to be difficult areas with special requirements and are as follows:

1   Mid-Levels, are areas of steep hillside covered with colluvium derived from landslip debris. Control is exercised over any form of excavation, site formation and foundation, including ground investigation in this area, PNAP APP-30 [45].

2 & 4   North-western New Territories and Ma On Shan are areas that are underlain by karst marble. Karst is difficult ground with respect to ground water in cavities and with respect to bearing capacity on cavity bearing rock, PNAP APP-61 [46].

3   Vicinity of Railways, to protect railways from the effects of construction. PNAP APP-24 [47].

5   Sewage tunnel protection areas, to protect DSD existing drainage and sewage tunnels. PNAP APP-62 [48].

Construction must be in accordance with Approved Plans. Construction must be carried out by an RC and must be under supervision by the AP, RSE and RGE accordingly. The number and qualifications of the site supervisory staff vary in accordance with the magnitude of the project and any special characteristics of the ground; they are stated in PNAP APP-28 [49].

Generally, AP, RSE and RGE are required to carry out periodic supervision. The RC is required to provide full-time supervision. For sites with geotechnical complexity including deep excavations, full-time site supervisions are required of professional grade geotechnical engineers under the periodic supervision of the RGE as stated in PNAP APP-48 [50]. If circumstances change from those assumed in the design requiring the design to be amended, a minor amendment can be annotated on the approved plans for approval by BA. A major change requires a new submission.

When a project is completed, As-built plans are submitted to BD, and an application is made to BD by AP for an OP. As-built plans are held by BD for future reference.

3.8 Planning the works

It goes without saying that design for deep excavations is necessarily predicated upon the type of ground. For designs to be workable, and not impossible to build, or better still for designs to be efficient, they should take into account the site situation, the nearby structures and the available methods of working to construct the deep excavation. A single deep excavation is often planned for a specific site, but some major projects start with site selection. For public works sometimes a site is to be found or may be selected within a given region. Site selection involves a number of factors including suitability of size and access and is sometimes selected based on ground conditions. For a large site, the disposition of construction within the site might be affected by the ground conditions. For some projects, the orientation of the layout might be adjusted. Access to the site is a factor, but usually it is not an over-riding one. In most private building works, the site is already defined by the title to the land and, come what may, the building works shall be carried out irrespective of how difficult the ground conditions are or how sensitive the environs are to construction work; the role of the designer is to conduct as thorough a site investigation as is feasible. Site investigation entails finding out information about the site, which includes terrain, topography, previous uses of the land and ground investigation. Ground investigation is an investigation of the types of soils and rock beneath the site by sampling and testing.

3.8.1 Desk study

At the start of a project a desk study is carried out; this means researching as much information that is published and that is held on record relating to previous projects carried out on the site itself and in the vicinity. The procedure often starts with information on the geology, seismicity and historical use of the site. A search should be made for any records, for example, any previous earth works, say for mining or filling over a previous landform including any buried structures such as former sea walls or breakwaters, in the case of reclamation from the sea or foundations and basements from a previous building on the site. Recent reclamation is not without problems. The underlying natural soils might be under-consolidated with positive pore pressures yet be dissipated and very weak, or the weak soils may have been disturbed, thereby reducing their strength if the soil is sensitive, or the soil may have been thoroughly disturbed and displaced, forming "mud waves" by the reclamation process with very little firm fill on top. As I recall, a novel method of construction of a deep excavation in very weak ground was adopted for Marina Bay Station in Singapore, where the depth of newly placed sand fill and underlying very soft marine clay was considerably more than the depth of the excavation. The under-consolidated marine clay would have provided no passive restraint whatsoever to the bottom portion of the diaphragm walls. The excavation at Marina Bay was kept full of water until the required depth was reached for concrete to be poured by tremie method (pouring the wet concrete through water, across the base of the excavation to form a slab to brace the diaphragm walls before the water is pumped out, as I recall). Archaeological remains can be a major issue, as work on the project may be halted for a long period or stopped altogether. Areas rich with archaeological remains, such as Athens, can require extensive activities before any construction work can take place. During construction, there is always a risk of discovering ancient artefacts. For example, baked mud ovens dating a thousand years or so were discovered on undeveloped land during bulk excavation for the Chek Lap Kok International Airport in Hong Kong in 1995[51].

In Hong Kong, the GEO established a Civil Engineering Library with over 164,000 documents [52] including a digitised database of archived borehole logs that now exceeds 400,000 boreholes. The database can be accessed on line. It supports spatial and textual searches for users to locate needed information efficiently. It provides a valuable source of information about the ground, especially in the built-up areas. The Singapore Government has recently started a similar database. Another source of information is an archive of construction record plans. For many years, the BA in Hong Kong has required that as-built plans for private building works, including records plans for foundations, be deposited [36]. Likewise, in Singapore the BCA has an archive of as-built plans for private building and for public works; they must be submitted at completion of a project [53].

When referring to archived information, one should bear in mind that some of the old records could be imprecise or even unreliable. A contractor was tunnelling below King's Road North Point for the Hong Kong Island Line railway tunnels; he relied upon record plans showing short piles driven to bedrock for an adjacent six-storey building. As I recall, when the tunnelling went past the building, it suddenly tilted; the building was temporarily vacated and a subsequent investigation revealed that the bedrock was well below what was indicated on the as-built records.

For information about utilities, enquiries should be addressed to all owners of utilities for their records of the vicinity. In some cases, utilities may have to be diverted before construction can begin, and this may have to be arranged by the owners of the utilities and possibly diversions should be carried out as advance works before the deep excavation begins. In some cases, utilities can be supported during the deep excavation. In areas of old development, the records of utilities might be not very accurate, and there might be unrecorded utilities. Often it is considered necessary to carry out trial pits or trenches to identify the locations and conditions of buried utilities.

3.8.2 Site-specific investigation

Having carried out a desk study, a site-specific investigation can be carried out by ground investigation using specialist ground-investigation contractors. The most common method of ground investigation for deep excavations is boring below the intended depth of the excavation to identify the type of ground down to the final formation level and to further depth to identify the type of ground that might be affected by the works or that needs to be identified for purposes of carrying out the design. Sometimes more borings are undertaken at the start of the construction contract in order to obtain more information about the ground (e.g. to finalise in detail depths for diaphragm walls or depths of hydraulic cut-off between the borings carried out for purposes of design). Borings on site can also include in-situ tests, and samples can be retrieved for identification and testing as described in Section 5.1.2. Some samples can be used to take specimens for testing in a laboratory. Boring records are kept for purposes of the design and for the planning of construction works. Samples are retained for later reference. They may be made available for tenderers to inspect and should be kept in good condition for reference in case claims arise relating to ground conditions.

For very large projects, all of the geological, relevant archived and site-specific data can be compiled in a geological model. The model is the understanding and interpretation of the available data. It is contained in a numerical database that can be accessed and interrogated with the Geographic Information System [54] and, more recently, BIM software [55]. For example, a geological model for a series of tunnels along the north shore of Hong Kong Island included reference to, and data from, some 35,000 boreholes, mostly archived [56]. For smaller projects, the geological model would not usually be so elaborate.

Site-specific ground investigation usually comprises a number of boreholes with in-situ testing, retrieval of samples and testing in a laboratory. Other techniques can be useful. Inevitably, there are gaps between boreholes; therefore, a large proportion of a site is unexplored. Determining the ground conditions between boreholes can be tricky. Profiling can be carried out by sensing. Methods include geophysical seismic surveys, electrical resistivity surveys, micro-gravity surveys and ground penetrating radar [57]. Each method is indirect and requires calibration with boreholes; therefore, it is recommended that the surveys be conducted along lines of boreholes for calibration and establish profiles between boreholes. Geophysical seismic surveys involve sending a percussive wave into the ground by striking a plate on the ground with a hammer or by a small detonation and recording the time the wave takes to reach a spread of "geophone" receivers. The results can be interpreted to estimate the depth to firmer ground below less firm ground (e.g. dense soils below soft soils and depth to rock). Profiling is achieved by moving the striker plate and the receivers along the line intended for the profile. With care and repeated intervals profiling (e.g. for top of firm soil, say SPT N values = 30; top of hard material, say SPT N values 100) rippable rock and massive rock are possible with skill. Electrical resistivity surveys are conducted by inserting two electrodes into the ground and measuring the resistivity between them; the electrodes are moved progressively along the line for profiling, and the results are interpreted in terms of resistivity versus depth. The resistivity changes significantly in passing from dry soil into wet soil, and below the water table changes in resistivity are more subtle. Micro-gravity surveys involve a single instrument to accurately measure the gravitational filed point by point along the line of the intended profile. The changes in gravity along the line are computed to represent changes in density at various depths. Changes in gravity are very small, and the survey is not very discriminating. Indications are obtained for depth to rock, pinnacles of rock head in soil and large caverns in karst. Ground penetrating radar makes use of devices for transmitting radar into the ground and receiving the returning signals. The technique has limited depth of penetration and is useful for locating utilities. Probably the most useful application of profiling for deep excavations is geophysical seismic surveying conducted along the proposed line for retaining walls. Profiling is then used to the occurrence of rock that is expected to be close to the founding levels of the walls. Profiling can also be used generally across the site if rock is expected during the bulk excavation. Generally, rock is more difficult, costly and time consuming than soil; profiling can help to determine the amount of rock to be excavated. Soil properties can be determined from geophysical profiling [58].

Ground investigation should also consider environmental concerns especially if the site is a brown-field site that has been previously used for an activity that could have polluted the ground. Boring and sampling for environmental studies have to be carried out without contaminating the samples and with methods that trap volatiles. Details for ground investigation for environmental studies can be found in national Standards [59, 60].

3.8.3 Procedures

The main elements of design are the structural details for the earth lateral support system. These comprise retaining walls around the perimeter of the deep excavation, bracing to restrain the retaining walls and details for both the hydraulic cut-off and instrumentation, including settling the AAA limits, as described in Section 3.4. In addition, there may be supporting works or underpinning required for nearby structures.

The procedures for design of public works and private works may differ. For public works, the Engineer is often the coordinating consultant. For private works, the Architect is often the coordinating consulting and the Geotechnical Engineers and the Structural Engineers are sub-consultants for the ground engineering works. The coordinating role concerns preparing the form of contract, liaising with third parties, obtaining necessary clearance and permits. Architectural forms of contract are intended for building works, which essentially involve processed or manufactured materials, excepting for foundations that are affected by the ground and geological risk. However, for a high-rise building, foundations can cost 10% or less of the total cost, and an over-expenditure of 50% for foundations would add only 5% to the total cost for construction. In architectural forms of contract, risk to the contractor is reduced when price adjustments are made for basic costs such as fuel oil, steel, cement and labour in accordance with published indices. For deep excavations, ground engineering comprises a large proportion of the costs; variations in the nature of the ground can have a major impact on the cost and time for construction. Accordingly, civil engineering forms of contract consider geological risk in several ways to clearly define the geological risk that is taken by an employer and how much is to be taken by a contractor as discussed in Section 3.6.5. It is important when the coordinator is an architect that the geological risks be clearly understood by the architect and employer and that the conditions relating to geological risk be clearly defined in the set of contract documents. Sometimes the underground works can be carried out as one contract with the allocation of geological risk defined and the building and architectural works can be carried out under a different form of contract.

Another difference in the procedure is in the extent of design prepared before award of contract. In Hong Kong and some other places, there is strict building control for private building works under the Buildings Ordinance [36] as discussed in Section 3.7 whereby detailed drawings have to be approved before consent is given for construction to commence. The registered AP is responsible for the coordinating works; the RSE is responsible for the design and supervision of structural works and the RGE is responsible for design and supervision of geotechnical works. For public works in Hong Kong, the design for deep excavations can be prepared with an engineer as the coordinating consultant. The design can be an outline for purposes of tendering, and a contractor is invited to produce its own detailed design. The design can also be an engineer's responsibility for build-only contracts, for which all general layout and details are produced for purposes of tendering and the detailed design drawings are finalised as the work progresses. Public building works are not controlled by the BA, but drawings for public building works are to be submitted to the BA for reference.

3.8.4 Select structural form

The selection of the structural form for the earth lateral support system is made according to the type of ground and ground water levels, the working space on site and the type of equipment available, as well as the end purpose of the deep excavation. In a developed area, restrictions on noise, vibration and dust apply. Heavy driving for piles might be ruled out because of the noise, vibrations and exhaust. In preparing a programme for the work, transporting plant and materials to the site and exporting excavated soil from the site should be taken into consideration since these can affect the rate of working; it is important that the contract period be sufficiently long to allow for the completion of the work.

Space within a site is also an important factor. For a large site with plenty of space outside the walls, almost any type of plant for diaphragm walls or piling could be used. Prior to bulk excavation, installing the retaining walls can be carried out inside the excavation area; the working space outside the walls need only be sufficient for the overhang of the equipment used for installing the wall. If the site boundary is immediately adjacent to another building, it is difficult to build a retaining wall in contact with the adjacent property. Steel sheet piling can be driven right on the boundary line by using an offset follower, but the thickness of the steel sheet piling, normally about 450mm, affects the location of the permanent wall. For most pile, boring equipment needs to be a space of about 600mm or more outside completed piles. Some diaphragm wall equipment can operate very close to an adjacent building without a guide wall on that side, but there is a risk that loose soil beneath the building will need support. Grouting beneath the adjacent building to stabilise the soil requires permission from the owner of the building, which might not be forthcoming. Diaphragm walls can be used as permanent walls; however, they range in thickness from 600mm to 1500mm. The thinner walls have considerably less bending capacity than the thicker ones and may not be suitable for excavations deeper than about 12m because of the need for very close bracing, unless the ground is sufficiently strong and the ground water table is low, or lowered, during the temporary stages of excavation. Installation of diaphragm walls requires a slurry treatment plant. When using grabs to excavate the slurry treatment plant comprises a mixer and a few storage tanks.

When using cutter machines for diaphragm walls, the slurry treatment plant is quite extensive and includes grilles, sieves, cyclones and a press for fine grained soils. Also, bunkers are needed to contain the recovered spoil, and large tanks are required for storing slurry. On small sites, the slurry treatment plant can occupy the whole of the centre of the site. On narrow sites, the plant might have to be moved during the work in order to provide working space for the cutting equipment, which extends between 5m and 10m from the diaphragm walls. Sometimes the slurry treatment plant is stacked on several levels in order to fit within confined sites. Designing temporary retaining walls and permanent walls for deep basements on small sites where space is at a premium is a tricky task. If there is no space for the diaphragm wall plant to operate on a site, then thinner piled walls may have to be used possibly including ground treatment to strengthen the ground. The net floor area is an important consideration for real estate developers, and designers are encouraged to make the combined thickness of temporary and permanent wall as thin as possible.

Once the retaining walls are completed, the remainder of the excavating and installation of bracing is not restricted directly by the immediately adjacent buildings. Selection of levels for bracing is almost procedural because many braces have been constructed with bracing levels at about 3m spacing in order to provide about 2.5m headroom for the excavation plant. Once bulk excavation has commenced there is no longer a working platform at ground level in the excavated area. Access to the site at ground level is generally required for vehicles to enter. Many vehicle entries and exits are required to export the excavated spoil unless special arrangements are made (e.g. with a conveyor belt to another loading area or in the case of a site alongside a sea wall or river wall to a vessel loading point). On large sites, such as the West Kowloon Station in Hong Kong, a site road on a ramp can be used for road-going vehicles to drive down to a lower level for delivery of plant and materials and, importantly, to removed excavated soil. For very small sites, the minimum requirement for space can be an overhead gantry resting on the retaining walls and spanning a shaft for mucking our excavated soil with a grab and loading the spoil onto a lorry to be exported. Site facilities, such as offices, can be located above the level of the gantry. Larger sites can have a decking at road level either temporary decking or permanent construction in order to provide working space for hoisting equipment such as a gantry of cranes and vehicle loading and turn-around areas as well as offices, stores and workshops.

For sites with working space outside the excavated area, there may be no need for any decking over the excavated area, as long as cranes parked alongside the retaining walls can reach sufficient locations of the excavated area to be able to hoist out excavated spoil efficiently and lower plant and materials. Alternatively, a gantry that is built spanning the excavated area can be used for purposes of hoisting spoil and lowering plant and materials and overhanging part of the site, or a small decking for loading trucks. For very large excavated areas the width may be too far for cranes to reach and decking is required for cranes to access most of the excavated area and for vehicles to reach the cranes to be loaded with excavated soil.

In Section 2.2, top-down and bottom-up sequencing of construction are described. These two methods of construction affect the design for temporary and permanent works. The bottom-up method with temporary earth retaining structures is the simpler to design. Using this method, the excavation is completed to the formation level before any of the permanent works are constructed. The permanent works can be constructed as if they are in the open except that the bracing restricts some of the working space. There is no structural interaction between the temporary structures and the permanent structures. If diaphragm walls are used for the earth lateral support system and for permanent walls, the design for the diaphragm walls should include all the stages of construction and the deflections resulting from the excavation sequence, which will be locked into the permanent structure. Importantly, the permanent structural connections between the diaphragm walls and the various floors have to be detailed. For normal construction with a series of pours of concrete, steel reinforcement starter bars are left protruding from the first pour by a lap length.

Steel reinforcement for the subsequent pour is fixed overlapping the protruding steel from the first pour. In the case of diaphragm walls as initial pours of concrete, steel reinforcement has to be overlapped with, or coupled to, the reinforcement for the several floors to be cast later. Many years ago, designs included full lap length starter bars for steel reinforcement, which was included in the cage of reinforcement for the diaphragm walls. After excavating and during the bottom-up construction the concrete cover to the starter bars was hacked off or excluded by attaching sheets of Styrofoam to the reinforcement cage for the diaphragm walls; starter bars were then exposed and bent out at right angles to the wall in order to lap with reinforcement for the floors. There were two problems with bent starter bars: many of the starter bars were damaged and reinforcement cages were not placed sufficiently accurately in elevation for the starter bars to line up exactly with the reinforcement for the floors. These two problems were addressed by adopting couplers so that starter bars of a full lap length were not required, and short lengths of reinforcement were coupled to the reinforcement for the floors. Sometimes proprietary couplers are drilled into place in the diaphragm walls, but care has to be taken to avoid damaging the steel reinforcement in the diaphragm walls.

When adopting the top-down sequence, parts of the permanent structure are built as the excavation is taken down in stages. The working method has to be planned. For example, in order to hoist spoil, openings in the permanent structure should be located where cranes or gantries will be located. Then a decision can be made about how much of the permanent structure is to be built on the way down or completed later. If diaphragm walls are used, then connections to them are made in the same way that connections are made to diaphragm walls using the bottom-up sequence as described in Section 2.2.

3.9 Design

Traditionally, geotechnical engineers investigated the ground conditions for the site and wrote a report; structural engineers received geotechnical reports, sometimes added a margin of safety, and adopted simplified approaches to determine loads on structures they would then design. Designs for earth lateral support and permanent structures were often conservative but were safe and robust. Interaction between geotechnical engineers and structural engineers was minimal, and interaction between the ground and a structure was not considered in any detail. During excavation, the ground moves, and the designer is required to keep such movements small and within safe limits. When the ground moves, the stresses in the ground change. For example, if a retaining wall deflects forwards towards an excavation, the loading on the wall from the retained ground reduces. The loading on the wall largely determines the deflection of the retaining structure, and there is an interaction between the ground and the structure.

During the last four decades, increasing use has been made of computers to aid designers. Interaction between soil and structures, especially during the excavation and construction stages, is easier to calculate with a computer than by hand. It has become increasingly important that designers of deep excavations be proficient in both geotechnics, to determine the characteristics of the ground and ground water, and in structural design and the interaction between the ground and structures.

Commercial computer programs now contain some very complicated numerical modelling of the behaviour of soil. For the design of deep excavations the modelling of the behaviour of soil in some of the powerful commercial computer programs is too refined for the inherent uncertainties of determining the types of soil, their extent and their engineering properties. Such programs are designed to be easy to use, and their sophistication can be overlooked. I have heard well-respected structural engineers say *"PLAXIS is a structural engineering program and should not be used by geotechnical engineers."* I doubt that Peter Vermeer and Helmut Schweiger, the originators of PLAXIS, would agree with that. My colleague geotechnical engineers frequently use PLAXIS to model stages of deep excavations from which they export the ground pressures as loadings for structural engineering colleagues to use as input data to structural engineering programs.

It is important that designers of deep excavations understand, and be proficient at, both geotechnical engineering and structural engineering.

3.9.1 Design standards

During the Planning Stage, as discussed in Section 3.8, the form and the sequence of construction have been determined in the preliminary design. During the Design Stage, design calculations and drawings are prepared. For purposes of tendering a contract, the engineer usually provides an outline of the design that is intended to be sufficient for the contractors to prepare a tender. For contracts that are let "build only," the engineer provides a detailed design. For "design and build" contracts, the engineer provides only the outline design as a reference design, and during the contract period the contractor provides the detailed design for construction. Design Standards should be adopted depending on the requirements of the authorities. In Hong Kong, British Standards and Codes of Practice were adopted or used as reference during the British rule prior to 1997 and are still in use. Steps are being taken to adopt Eurocodes, which have been adopted in a number of countries.

Fundamentally, because the properties of the ground cannot be determined exactly under every circumstance, design is based on a form of risk assessment and mitigation. The design is intended to be robust, so that under a wide range of anticipated ground conditions the excavation, the structures in the excavation and surrounding area remain serviceable and if extreme conditions are encountered, a structural failure or a total collapse does not occur. Moreover, the works should be monitored, including deformation of the structures and movements of the ground. Monitoring should confirm that everything is going as planned. If the monitoring results indicate design assumptions might not be met, for example the settlement of the nearby ground is developing more than expected, preventative actions shall be taken.

Designing for robustness is achieved by two methods, deterministic design and limit state design. Deterministic design assesses the typical values for the salient engineering properties of the ground and uses the values for calculations of stability with a prescribed factor of safety.

Typical values are determined at the discretion of the designer, from the average, or close to average, values from a large number of tests, and feedback from works of a similar nature in similar ground. If less data is available, the values that are adopted for design should be more conservative. The stability of elements of the works is assessed and factors of safety are applied. Perhaps the simplest example is a retaining wall with a rough base cast on soil such that sliding of the wall is resisted by the shearing strength of the soil below the base. In the analysis, the shearing force that is required to prevent the wall from moving is calculated. A factor of safety of 1.5 means that shear strength of the soil that is in contact with the base of the wall is 1.5 times greater than the calculated shearing force. If, in an extreme event, the strength of the soil below the base happens to be two-thirds of the assumed strength, the factor of safety would be 1.0. That means that the applied load on the wall from the retained ground would be about to overcome the shear strength of the soil under the base of the wall and sliding failure of the wall would occur. When adopting a deterministic method of design, the designer is not required to assess what extreme cases might be. Codes of practice specify values for factors of safety to be adopted on the basis that a typical strength is assumed.

Limit state design is embodied in Eurocode 7 [29], which recommends the use of partial factors and evaluation of both the serviceability state and ultimate limit state. The fundamental principle is similar to the deterministic approach, but instead of one overall factor of safety, a partial factor is applied to the material strength, such as the shearing strength of the ground beneath the base of a retaining wall; a partial factor is applied to increase the loads. Both the serviceability state and the ultimate limit state are checked using specified partial factors. In addition to the partial material factor, Eurocode 7 recommends that moderately conservative rather than typical soil parameters be chosen and adopted. "Moderately conservative" suggests a value that is more cautious than typical and implies that the designer shall allow for uncertainty in the derivation of the parameters beyond the recommended partial material factors.

If the deterministic approach and the partial factor approach are compared for a simple structure, for example a strut, the results can be comparable. If the partial factor on material strength is multiplied by the partial factor for the loads, one gets an overall factor of safety. For example, if a material factor of 1.2 is adopted and a load factor of 1.25 is applied, their product of 1.5 is an overall factor of safety. For embedded retaining walls, there is a significant difference. Below excavation level, there is a point of zero load where the passive pressure on the embedded inside face of the wall balances the active pressure on the back of the wall as shown in the Figure 3.3. The net load on the wall is the thin hatched triangle with above the point of zero load.

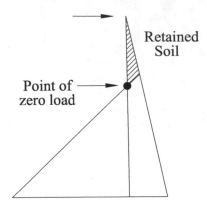

Figure 3.3 Un-factored strength.

If the active pressure is considered to be the load on the wall and is factored higher and the passive pressure is considered to be the material that is loaded and subjected to a material factor and is factored lower, the point of zero load is lower as shown in Figure 3.4 and the height of the loaded section of the wall, shown hatched, is increased. Thus, not only is the intensity of the loading increased but also the loaded depth is increased.

Therefore, the total load on the wall is factored twice, once for the partial load factor and again because of the increased depth due to the difference between the factored load and the factored resistance. When adopting a deterministic method, the loads described in Figure 3.3 are applied and factored by the factor of safety. The maximum bending moment in the

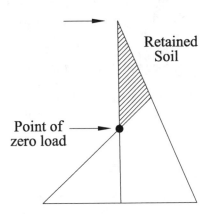

Figure 3.4 Factored strength.

wall is then located at about the excavation level. By contrast when partial factors are adopted the loads described in Figure 3.4 are applied, the maximum bending moment in the wall is located deeper, below the excavation level. The deterministic approach, in effect, strengthens the location of the typical maximum bending moment but the partial load approach strengthens the wall where the maximum bending moment would be if the soil happens to be weaker than expected. The partial load factor method seems the more logical approach since the strength of the soil is the variable parameter at risk.

Analysis of the interaction of the soil and the structures by numerical modelling is used to determine the required structural capacity of the earth lateral support system comprising the bracing and the retaining walls. Therefore, it is now common to carry out numerical modelling with a computer program, with moderately conservative soil properties and partial factors applied to the soil strength and to the applied loads. The output is then used to determine sizes of structural members in accordance with limit state code requirements. However, the properties of soil are moderately conservative, which is generally assumed less than the typical strength and a partial factor, which reduces the strength further to a value that has a lower probability of being encountered. For example, if the average strength obtained from many tests is 100kPa, and assuming for purposes of illustration that strength measurements have a normal distribution, and that the standard deviation of the results is 10kPa then the mean minus two standard deviations is 80kPa. At two standard deviations the probability of occurrence of a lesser value than 80kPa is 2.5%, which could be considered a moderate or a very conservative value depending on the circumstances, such as number of tests and range of different soil types identified and tested. Applying a partial factor of 1.2 to the strength of soil lowers the strength to 66.7kPa, which is two-thirds of the mean value. This value is 3.3 standard deviations below the mean, and the probability of occurrence of a lesser value is about 0.1%, which is 1:1,000. The stiffness of soils is related to strength. If the stiffness is lowered to two-thirds the average value, then the estimated displacements will be considerably more. For elastic materials, the displacements would be over-estimated by 50%. After the onset of plasticity, the reduced strength becomes effective, and the estimated displacement can be over-estimated considerably more than by 50%. It is prudent to ensure a margin of safety in the capacity of the earth lateral support system compared to what is to be expected typically, applying a partial factor that substantially overestimates displacements that would affect the surrounding area and could be considered unnecessarily conservative, especially since instrumentation and monitoring are adopted during construction to check on the displacements that actually take place. It follows that for many projects two numerical models are carried out, one with partial factors for compliance with the design code and one with deterministic parameters perhaps representing 90%, the majority, of the tested range of soil properties; the resulting displacements are a working estimate that can be used for example in determining the likely effect on nearby buildings and in setting the AAA limits. Factors of safety are not just required to manage risk of unexpectedly weak soils; they are also intended to provide reserve capacity in the system for other unforeseen events such as accidental damage (e.g. on construction sites with plant running into or knocking temporary structures or over-excavation of which only half a meter results in a substantial additional load on the system until the next level of struts are in place, accidental flooding and so on).

The Deterministic Design using typical values for parameters leads to estimates of expected performance, for example, for expected bending of a retaining wall and for expected settlement of the nearby ground. The Limit State method applies factors; therefore, the results of an analysis are estimates of performance under conditions more onerous than expected. There is a margin put into the calculations by adopting partial factors. For purposes of determining the structural capacity or for stability, partial factors are used. For estimating expected performance (e.g. for comparison with monitoring results), unfactored analyses using the deterministic method are required.

3.9.2 Design considerations

Before calculating the number and sizes of structural members, there are several factors for the designer to consider. Have the types of soil throughout the extent of the proposed excavation and for an adequate depth below been investigated? Has the investigation been detailed enough? Have enough samples been taken and enough tests carried out? Are the test results consistent with available data for the same types of soil in the vicinity? Has the ground water been studied sufficiently? What are the seasonal variations? What is the chemical content of the ground water?

What is the site area? What types of equipment are available? Is the working area sufficient to use them? What headroom do they require? Are there suppliers of ready-made struts for hire? What is located nearby? Are there utilities and structures that might be affected? What does the employer require? What type of structure? What degree of water-proofing ? How close must the retaining walls be to the boundaries?

The designer is then in a position to consider, and then decide on, what methods of construction will be used, what types of earth lateral support will be adopted. In some cases, alternative methods and types of construction may be considered to the stage of an outline design for purposes of evaluating cost and time so that an informed decision can be made as to which to use. Very often diaphragm walls are adopted, and they can be used for temporary earth retaining walls or for both temporary and permanent use. The choice between top-down sequence and bottom-up sequence of construction affects the design and needs to be decided at an early stage.

Bracing is necessary to support the retaining walls and limit their movements to control the subsidence outside the excavation. Steel struts commonly used as bracing shorten when it is loaded for two reasons. One is looseness in assembly and the other is elastic compression due to the axial loads. Pre-loading is generally adopted for both purposes. Light preloading can take up the slack on assembly; this is important to ensure that the various components including shims and bearing plates are tightly in position and do not fall out. Heavier preloading, nearly up to the predicted maximum load, can compress the struts by the elastic strain, which can be of the order of 15mm to 20mm for a 20m long strut and reduce the amount of deflection of the wall accordingly.

For relatively narrow deep excavations, it is usually the case that several levels of struts at about 3m headroom for excavation plant is more cost effective than fewer levels of struts with a higher capacity and walls with a higher structural capacity. For wide excavations, the choice of bracing system is less obvious. In the case of West Kowloon Station [21], which is 200m wide, the construction was let as two packages. Both contractors commenced work by constructing diaphragm walls around the perimeter. The contractor for the southern half opted for top-down construction whereby most of structures for the six floors below ground were constructed top-down on the excavated surface at each stage and as the excavation progressed, each stage was mined beneath the permanent floor above. The contractor for the northern half decided to excavate more quickly all the way to the bottom in the middle of the site leaving unexcavated berms of soil to support the diaphragm walls. The middle portion of the structure was constructed bottom up, casting columns and floors conventionally using formwork. The outer portions of the structures were constructed sequentially top down as the berms were excavated stage by stage and struts were used between the diaphragm walls and the already completed structure in the middle portion of the site. Both contracts were let by competitive tender; the nature and extent of the work was similar, but two different methods were chosen by the contractors.

By adopting two fundamentally different sequences for construction, there were several interfaces between the two contracts. For example, the diaphragm walls deflected differently. In the south site, the top-down method resulted in the walls progressively bending inwards with the maximum deflection near the bottom of the excavation being restrained from the top by a succession of floors. In the north site, the diaphragm walls acted as supported cantilevers and the largest deflections were at the top of the walls until the later stages of construction. Therefore, the adjacent diaphragm wall panels at the boundary between the two sites were subjected to warping. Because the vertical joints between diaphragm wall panels can rotate about a vertical axis, the warping only extended to about three panels to either side of the boundary and had little structural effect.

3.9.3 Structural members

When designing the structural members for a deep excavation, there are several areas for optimisation. For example, let us consider the struts that are used for bracing. The cheapest way to procure steel struts with a certain cross-sectional area is to use rolled sections, straight from the catalogue of a steel manufacturer. However, such steel can be re-used and hiring ready-made struts can be economical because of the re-use. Using ready-made struts can also be quick because even the minimum amount of fabrication such as welding on end plates for bolted connections has already been done by the supplier. It should be remembered that rolled steel sections are available only up to a size of about 90cm, which compared to a span across the width of a braced excavation of about 18 metres, means that the strut is very slender; to develop most of the strength of the steel, the strut may have to be braced with vertical posts and laterally with ties from one post to another. Steel members of larger sizes are available as pipes or can be assembled by joining two or more standard sections together to act compositely. Fabrication of special struts of unusual shape requiring a lot of fabrication should be used only under special circumstances.

For reinforced concrete structures, optimisation is a question of whether to have thin cross-sections with high percentage of steel reinforcement or thick cross-sections of concrete with less steel reinforcement. As a general point, underground structures extending well below the ground water are buoyant, and weight is needed to resist buoyancy. Therefore, concrete structures that are thick add to the stability and benefit from reduced steel reinforcement. When making concrete, using especially heavy aggregates including shards of scrap steel is an expensive way to resist buoyancy and is usually not adopted. For internal suspended floors, minimum weight might be an advantage, and there are several choices of structural forms for internal floors as there are for buildings that are above ground.

When reinforced concrete is used as bracing to support retaining walls, it should be remembered that concrete shrinks after casting. This effect is the opposite of preloading that can be adopted when installing pre-fabricated struts because the retaining wall deflects by an amount equal to the shrinkage of the concrete bracing before the concrete bracing develops support to the wall. For large spans of concrete, the shrinkage can be several centimetres. The effects of shrinkage can be reduced by leaving an open strip, a "pour strip," alongside the retaining wall, pouring the bulk of the concrete first, then allowing it to shrink and then casting the pour strip. By this means, shrinkage is accommodated before support to the wall is developed; there is only shrinkage in the narrow pour strip. Pre-load can be applied by placing jacks within the pour strip before concreting them. When working underground there is merit in keeping procedures simple "KIS" and, in this case, not pre-loading a pour strip.

3.9.4 Slopes and berms

Slopes and berms that are sometimes adopted as part of the earth lateral support system should be checked as passive support with a down slope in front of the retaining wall. It is possible to use slope stability programs by including the embedded part of the wall with a lateral thrust imposed on it. Failure mechanisms are usually a sliding surface or surfaces from the toe of the wall or part way down the embedded part of the wall going upwards through the berm, Figure 3.5. An analysis using a slope stability program without the applied forces from the embedded part of the wall results in a slope stability failure with a sliding surface or surfaces of a different form commencing at the toe of the berm and passing up to the top of the berm at the face of the wall as shown diagrammatically in Figure 3.6.

3.9.5 Soil to wall contact

A technical consideration is the shear stress between the retaining wall and the soil. Computer programs allow the input of slip elements that can be weaker than the soil and can be used to represent adhesion or reduced friction between a wall and the adjacent soil. When diaphragm walls or piles bored with bentonite slurry are adopted, concerns are for the effect of bentonite on the friction or adhesion. Load tests on piles have resulted in a range of results from very low adhesion that has been blamed on a gooey skin resulting from contaminated bentonite to full strength of the soil as discussed above. Some of the guidance on the subject is to assign a slip layer with adhesion that is 80% or less than the shear strength, or

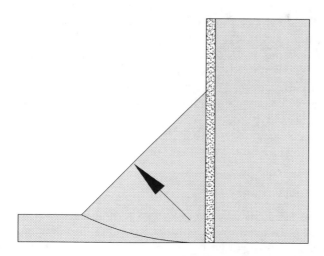

Figure 3.5 Toe pushes forwards.

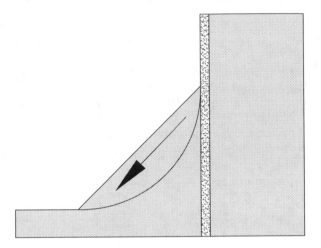

Figure 3.6 Slope slides down.

with an angle of friction of ⅔ times the angle of friction for the retained soil, ϕ' [61]. However, from my examination of several computer programmes, I observed that the vertical surface of the back of a wall was a plane of maximum shear stresses for the retained soil and not a plane of maximum obliquity. For soil with an angle of friction ϕ', on a stress plot, as shown in Figure 3.7, maximum obliquity is when the shear stress is the normal stress times tan ϕ', whereas the maximum shear stress is at is the normal stress times tan β, which is an angle that is about ¾ ϕ'. By adopting an angle of friction on the surface of ⅔ ϕ', one ends up with an effective friction of ¾ × ⅔ ϕ', which is ½ ϕ'. In practice, a reduced shear strength on the surface of a diaphragm wall could result from a "cake" of contaminated bentonite slurry, and a slip layer could be adopted in the analysis for that reason. Adopting a reduced angle of friction for a clean contact between a wall and a frictional material appears to be unnecessary when using a continuum numerical analysis.

Single panels of diaphragm walling are often used instead of bored piles as foundations. Several studies have been done of the shaft friction that can develop on the vertical faces of barrettes. For reliability and improved shaft friction, grouting around the perimeter of barrettes is often adopted [62].

3.9.6 Bulk excavation

Excavation in soil is usually done with mechanical excavating plant. The plant has to operate within the headroom below the bracing. Sometimes rock is encountered towards the bottom of deep excavations. Mechanical breaking of rock can be time-consuming. Blasting is instantaneous and

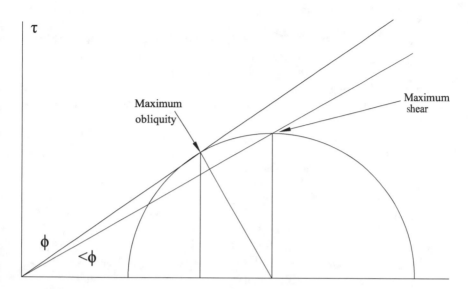

Figure 3.7 Maximum shear and maximum obliquity.

permissible if the blasting is controlled with small charges per delay and continuous monitoring. Blasting usually requires protective measures such as heavy mats to prevent fly rock. A number of chemicals, when placed in holes drilled in rock, will expand and split the rock silently and with no vibration of the ground.

Over-excavation can happen accidentally or loss of passive resistance at the level of the excavation may be caused by unwanted water. Therefore, it is prudent to design for a certain amount of accidental over-excavation in order to make the retaining wall more robust.

3.9.7 Retaining walls

The basic theory of design of retaining walls is described in Section 2.3. A retaining wall is a critical element of earth lateral support, and its failure could result in a collapse. The wall has to be robust, and usually there is a requirement to check the capacity of a wall to resist the effects of the failure of one strut. Failure of one strut would require a wall to span between the adjoining struts in the absence of the failed strut. After the collapse of the Nicoll Highway in Singapore as described in Section 5.2.3, the requirement for one strut failure was changed to failure of a whole row of struts as would be modelled in a plane cross-section of walls and bracing as one strut failure with no lateral redistribution.

When retaining walls are incorporated into permanent works, structural connections are needed from the walls to the future slabs or beams. Historically, steel reinforcing starter bars were provided within the cage of steel reinforcement for the walls and were exposed when needed and bent out. Often Styrofoam sheets were placed behind the starter bars to facilitate exposure and bending. An alternative now commonly in use is to have short starter bars for mechanical couplers. Problems can arise when locating a steel reinforcement cage below ground, typically within the slurry-filled trench for a diaphragm wall, and often the couplers and starter bars ended up out of tolerance.

Often a retaining wall of an underground structure has to be broken out in places to construct access. For connecting to pedestrian subways, the subway is often built up against the outside of the retaining wall and the retaining wall is cut through from both sides. When tunnels are built up to an end wall of a station with a Tunnel Boring Machine (TBM), it is often necessary to cut through the retaining wall using the TBM. Cutting steel reinforcement using a TBM is not easy, and a useful alternative is Glass Fibre Reinforced Plastic (GFRP) reinforcement. A section of walling with GFRP is referred to as a "Soft Eye." GFRP bars have some typical material behaviours that distinguish them from the more commonly used steel bars. First and most significant is the lack of yielding of the material. GFRP is linearly elastic until failure, which is brittle. Brittle failure behaviour allows a TBM to drill through the concrete and the reinforcement bars. Steel reinforcements would obstruct the TBM due to its yielding behaviour. When someone cuts through steel reinforcement, short lengths of distorted steel bars get tangled inside the TBM and block the passageways for removal of excavated spoil. Cutting through an end wall of a station by a TBM is usually required so that a seal can be installed around the end of the tunnel lining in order to prevent ingress of ground water or to provide a movement joint in seismic areas.

3.9.8 Seismicity

3.9.8.1 Seismicity of the site

The surface of the earth can be described as tectonic plates of relatively intact crust with faults around them. The earth's crust moves. Earthquakes happen in the earth's crust when built up stresses exceed the strength of a rupture zone in a fault; the rupture displaces with a release of energy. The release of energy sends a shock wave through the rock and through the soil where it occurs at the surface. The plates are quite large, such as the Pacific Plate underneath the Pacific Ocean, which is responsible for earthquakes in Taiwan and Japan, and is edged by the major San Andreas Fault in California. Because earthquakes happen at specific zones, their effects vary considerably from one location to another. In each country, the authorities classify zones of various intensities, and codes of practice make use of the broad classification of such seismic zones to prescribe intensities for the design of buildings [63]. The zones and codes of practice are intentionally conservative. For major projects, the seismicity of the site is normally determined more specifically. In some countries, guidelines are specified for determining site-specific ground motions for the design of major facilities. For example, guidelines are published for electric power plants in U.S.A. [64].

In principle, seismicity of a site is determined probabilistically from historical records of earthquake events recording the magnitude of the event and its location, the epicentre. Historical descriptive records are available from 1,000 years ago. Some historical records are imprecise. Gradually, the information became more specific and precise. Richter proposed a scale of magnitude of earthquakes in 1935 [65], which is basis of what is used today. Historical records are published to assist designers in determining site-specific design basis ground motions [66].

An introduction to earthquake engineering and seismic codes in the world is summarised by Ishiyama [67]. Earthquakes are defined by the epicentre, which is the location of the ground motion including the depth. The size of an earthquake is given by magnitude (a measure of the energy released at the epicentre) and intensity (a measure of the ground motion experienced by a receiver).

Precise instruments that measure the vibrations of the rock are called seismometers. They came into use in the twentieth century. By comparing the vibration records of the same event at several seismometers and noting the different time taken for the arrival of the seismic wave and its intensity at each location, one can compute fairly accurately the magnitude of the event, its location, in eastings and northings, and its depth.

Seismicity analysis relies on historical records and computing the intensity of the ground motion at a specific site. The ground motion reduces with distance from an epicentre, and the distance from the epicentre to the subject site should be calculated for each recorded event. The total is subjected to a statistical analysis to determine probability of re-occurrence. The frequency of exceedance of magnitudes of ground motion computed for the subject site is determined from as many events as possible, and a relationship between magnitude and frequency is thereby determined.

GEO has collected and published data on earthquake events that occurred within 500km of Hong Kong [66]. The data provide information for seismic evaluation for sites in the region.

3.9.8.2 Requirements for design

Earthquakes have several potentially damaging effects. One effect is liquefaction of soils with dramatic loss of bearing pressure, with subsidence, and with sand and water being emitted at the ground surface. Another effect is lateral spreading. A third effect is shaking: rapid ground motion that imposes distortion and inertial loading on underground structures. The effects of seismicity on deep excavations and the completed underground structures contained in deep excavation must be considered in seismic areas. For buildings works the magnitudes of seismic events are usually specified by the relevant local authority. Above-ground structures respond dynamically to ground motions, especially tall flexible buildings that vibrate more dramatically than their foundations and there is an amplification factor for this effect. For underground structures, the effects are less dramatic. Interaction of the buried structure with the surrounding ground means that the motion of a buried structure is equal to or less than the free field motion of the ground. Furthermore, in soil there is an amplification factor, as discussed below, whereby the deeper the excavation the less the magnitude of ground motion to which it is subjected. Building codes are generally conservative; for large projects, site-specific studies may be carried out to determine the seismicity of the site and appropriate parameters to be used for the design. For buildings, there is an importance factor that depends on the consequence of out of service after an earthquake or failure in the case of a major earthquake. For example, nuclear power plants have a very high importance factor because of the horrific consequences of massive failure. Some authorities and owners require high standards for the design of their facilities [68]. The San Francisco Bay Area Rapid Transit (BART) is a case in point. The 1989 Loma Prieta earthquake [69] had a severe impact on the transportation subsystem above ground in the Bay Area. Collapse of the Nimitz Freeway Viaduct alone resulted in 40 fatalities, and 3,757 people were reported injured. Major disruptions were widespread including the San Francisco-Oakland Bay Bridge and many parts of the highway system.

The 100km BART system remained operating because of the high standards of earthquake resistance adopted for the design. Without BART operating, much of the city would have been impassable. Designing underground structures for earthquakes renders them robust for other untoward events. In locations that are less prone to earthquakes, design standards sometimes include civil defence requirements such as resistance to damage from bombing. Underground structures than can be accessible to people are often planned as refuges and bunkers. BART remained operating during and after the 1989 Loma Prieta earthquake with magnitude M7.1. It may be noted that San Francisco is a high-magnitude earthquake area. If earthquakes struck randomly over time, the region would expect another earthquake of this same magnitude in the next 30 years with about a 50% probability [70].

3.9.8.3 Standards for design

There are usually two levels to consider, a lower magnitude event with a return period of, say, 1 in 100 years after which the structures shall be fully serviceable and a more extreme and less frequent event, such as 1 in 1,000 years after which the structures shall not have collapsed beyond repair. Standards for design vary according to the type of facility and the requirements of the owner. For underground railway systems in moderate-to-high seismic areas such as Taipei, Seattle and Los Angeles, the approach is to ensure functional adequacy and economy while reducing life- threatening failure. Two design events are termed as The Operating Design Earthquake (ODE) and the Maximum Design Earthquake (MDE). For underground railway structures, a typical definition of ODE is an earthquake that can be reasonably expected to occur at least once during the lifetime of the structure. In layman's terms, this would happen "once in one hundred years." The requirement is that the whole system shall continue operating during and after an ODE with little or no damage. In other words, withstanding the effects of an ODE is a serviceability requirement. The MDE is defined as having a finite but small probability of occurrence, say 5%, during the life of the structure. The requirement is that public safety shall be maintained during and after an MDE. In simple terms, some superficial damage such as cracking is permitted but there shall be no collapse. People inside underground railway structures should remain safe, and people above underground should not be threatened by the performance of the structure: no collapse and subsidence of the road above a station. There are cases of more stringent requirements for highly important facilities, for example that it is still be operable after an MDE. Guidance on analysis and design of buried structures can be found in NCHRP Report 61 [71].

3.9.8.4 How the ground moves

At the surface of the earth, an earthquake results in compressive waves and shear waves. The earthquake is quantified by two scales. The Richter Scale depends upon the amount of energy released at the epicentre, the location of the movement in the earth's crust. The Modified Mercalli system relates to the intensity of the motion at the surface. Fundamentally, the effect of an earthquake in a given location is ground motion in bedrock. In soil, the shear wave in bedrock is amplified as it passes upwards through soil. This can be simulated for oneself by shaking a plate with a table jelly and watching the top wobble more than the bottom. Soil is not usually as weak as jelly, but deep deposits of weak soils can more than double the magnitude of the seismic shear motion in the bedrock. Whereas in rock the motion is most severe at the depth of the epicentre and reduces at shallower depths up to the bedrock surface, soil exhibits more motion at the surface than at depth.

Ground shaking is a period of vibrations lasting perhaps several seconds in the main shock and after-shocks. The vibrations include several frequencies, just as noise is not a single note and includes fundamental frequencies of about 1Hz and superimposed higher frequencies. Most of the high acceleration that induces inertial effects results from frequencies in the range of 1Hz to 10Hz.

3.9.8.5 How do structures and deep excavations respond?

Tall buildings and bridges respond dynamically to the waves of an earthquake, and unless they are damped some structures can vibrate to a greater amplitude that the ground motion that induces the response. On the contrary, underground structures do not usually exhibit a significant dynamic response; they are subjected to deformation that arises from the ground motion and some inertial loading. It is generally sufficient to represent the effects of earthquake as if it is a statically imposed deformation or as an equivalent additional body force from the surrounding ground that is in motion. Underground structures such as bracing for earth lateral support and permanent underground structures are deformed by the incident ground motion. Both compression waves and shear waves that are incident upon an underground structure are greater at the surface than lower down and induce racking as shown diagrammatically in Figure 3.8. The racking arises from the attenuation of the shear wave and imposes tilting of vertical elements while horizontal surfaces in the ground remain plane. It is the top of the wobble of the jelly on the plate. Tilting applies to the retaining walls. Temporary bracing is relatively

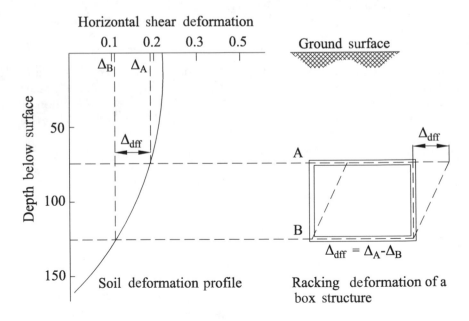

Figure 3.8 Free-field racking deformation imposed on a buried rectangular frame.

flexible, and racking can have little effect on the loads in the system; the design then can consider only inertial reactions from the ground motion in the retained soil. However, if there is structural connection between walls and horizontal structures such as floor slabs, the structure will rack and develop bending moments at the connections. The effects can be modelled as statically imposed deflections either directly on the underground structure adopting "greenfield" motions of the ground, or interaction between the buried structure and surrounding soil can be modelled by imposing the deformation on the boundaries of the surrounding soil. Alternatively, the effects can be modelled dynamically including inertial effects.

Seismic shear waves are variable not only in time but also in space. At any moment in time during the earthquake, the ground is deformed as a series of waves. When something straight is distorted into a curve, it gets shorter. This is sinusoidal shortening. For example, an underground pipeline that is deformed into a wave shape is pulled out of the manholes or junction boxes, and a railway tunnel is pulled out of underground station end walls. If tunnels have loose longitudinal connections around circumferential joints, the lining is flexible with respect to sinusoidal shortening. However, when longitudinal bolts remain in place after erection and if they are tightened, then the circumferential joints become stiff and the tunnels are shortened during the passage of a shear wave. For such long linear structures, flexibility such as movement joints should be included in the design. For large excavations, not only should waveforms be imposed on the structural system but detailing of connections to accommodate movements is also important.

When a dynamic numerical analysis is conducted, a seismic event has to be modelled including a realistic time history. A classic example is the El Centro Earthquake on 18th May 1940 because it was the first major earthquake to be recorded by a strong-motion seismograph located next to a fault rupture. The earthquake was characterised as a typical moderate-sized destructive event with a complex energy release signature. It was the strongest recorded earthquake to hit the Imperial Valley; it caused widespread damage to irrigation systems and led to the deaths of nine people. It was a large event with a moment magnitude 6.9 with available data at a time when dynamic analyses were being developed, Earthquake History [70]. Now time history records are available for several other large earthquakes of different sources.

3.9.8.6 Effects of earthquakes on slopes

Deep excavations in ground that is self-supporting can be formed with exposed slopes, and sometimes benches of soil are formed with sloping faces. The effects of an earthquake on a slope are inertial forces arising from accelerations during the ground motion and liquefaction in susceptible soils. The inertial effects are considered by two methods. For first is limit equilibrium methods, the peak horizontal acceleration force for the design seismic event is applied as a sustained equivalent pseudo-static body force. Seismic loads are added to a conventional limit equilibrium analysis to calculate a factor of safety. However, the ground motion is a vibration; it is not sustained. The peak acceleration may be sustained several times frequently but for only about one one-hundredth of a second before it is reversed. The second method for design that is now widely adopted is displacement-based analyses, which takes into account the short but frequent periods of high acceleration during an earthquake. There is a simplified method called the Newmark sliding block concept, and more rigorous numerical models can be used. This is a specialised topic, and a more detailed account may be found in the NCHRP Report 61 [71].

3.9.8.7 Liquefaction

Liquefaction occurs in loose soils below the water table, which, when shaken, would densify, but when the shaking is rapid as during an earthquake, and there is not enough time for the water to escape the excess water pressure builds up and the soil can flow like water. When slopes are exposed in a deep excavation, the ground water conditions are such that the slopes are not generally fully saturated under a discreet water table. However, ground water outside an impermeable retaining wall around the perimeter of a deep excavation often is little disturbed, and any susceptible soils retained by an ELS system could be subjected to liquefaction. Liquefaction generates excess water pressure, almost zero strength in the soils, and increased lateral earth pressure on the ELS. For further insight about liquefaction, the reader might refer to Idriss and Boulanger [72].

3.9.9 Contract period

For all types of construction, time for completion is important. Whereas government departments as employers for civil engineering contracts tend to take a realistic approach to allowing sufficient time for construction, sometimes politics intervene, and unrealistic targets are set. Private developers, especially for large projects with a lot of investment upfront, are usually intent upon completing the project as soon as possible and sometimes set unrealistically short times for completion. Putting pressure on contractors to complete early risks shortcuts or poor-quality workmanship and accelerating a programme by adopting unusual methods can prove to be very expensive. In some cases, it is simply not possible to go any faster. As a duty of care, an engineer who has evaluated a properly prepared programme for the works and realises that the time set by an employer is unrealistic, should make it clear to the employer at the outset and not award a contract when the contractor is likely to get into difficulties and over-run the contract period.

3.9.10 Tips about detailing

Attention to details is very important in structural design. It is sometimes said "the devil is in the detail." Studies of structural failures indicate that main structural members seldom fail. When failure does occur, it often originates from poor or inadequate detailing. If I recall correctly, the collapse of the first span of the box girder deck of Westgate Bridge, Melbourne, in 1970 was because relatively small but important "K-plate" temporary braces were inadequate, some of the steel plates buckled and matters got worse. The two halves of the first span plus some of the second span did not fit together and, whilst they were being weighted and were being jacked together, they totally collapsed into an occupied canteen killing 35 workers [73].

The collapse of the deep excavation adjacent to the Nicoll Highway in Singapore [74], as discussed in Section 5.2.3, was initiated by sway buckling of the connection of a steel strut to a steel beam waling that had no gusset plates at the ninth level of the excavation. The failure of this type of joint at the ninth level of strutting shed load to the eighth level of strutting where the same detail had been used. That too failed and total collapse ensued. For many decades gusset plates have been, and should be, a standard detail for connecting two steel beams at right angles [75] as shown in Figure 3.9.

In the interest of efficiency, details are often available from a library of standard details. Having standard details runs the risk that the details are adopted without checking. A case in point relates to the collapse of the deep excavation adjacent to the Nicoll Highway because just four years later in another city in South East Asia a very similar detail for the connection between struts and waling was used to support diaphragm walls. There were no gusset plates on the waling. The detail was adopted for all ten levels of steel struts as I recall. That retaining wall also collapsed

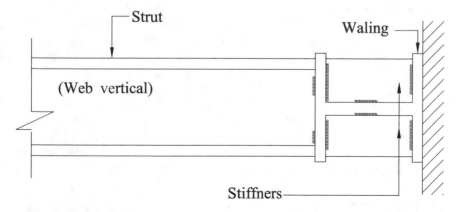

Figure 3.9 Standard detail of joint.

catastrophically, fortunately with no fatalities. I do not know the cause of this collapse.

3.10 Codes of practice

3.10.1 Standards and codes of practice

There are national standards and codes of practice for the design of earth-retaining structures. The European standard for design of geotechnical structures, Eurocode 7 [29], describes how to design geotechnical structures using the limit state design philosophy. Geotechnical structures include earth-retaining structures as well as ground-improvement work and foundations. Eurocode 7 addresses design and ground investigation and testing. Determination of geotechnical parameters is included as an integral part of geotechnical design. Eurocode 7 became mandatory for use in member European states in March 2010. Notwithstanding mandatory use of Eurocode 7 in Europe, a code of practice takes the form of guidance and recommendations. Compliance with a code of practice does not of itself confer immunity from legal obligations. Liability for the design of an earth retaining structure rests with the designer. In Great Britain, "*The Civil Engineering Code of Practice No.2*" (CP2) concerned the design of earth retaining structures and was issued by the Institution of Civil Engineers in 1951 [76]. The basis of the code of practice was the deterministic assessment of loading and geotechnical parameters and the application of overall factors of safety. The British Standards Institution completely revised this code in the late 1980s and published a more comprehensive code "*Code of Practice for Earth Retaining Structures*" in 1994 [77]. The revisions included the introduction of effective stress analysis for the assessment of earth pressures, whilst admitting total stress analysis for some walls during or immediately after construction. Also included was a recommendation to take account of movement, or lack of movement, on the resulting earth pressures on the wall. It was considered that the largest earth pressures acting on a retaining wall occur during working conditions and that these earth pressures do not increase if the wall deforms sufficiently to approach failure conditions. The code of practice takes into account that for small movements of a wall the shear strength developed in the soil is less than the maximum shear strength measured in a conventional triaxial test and furthermore that when large strains occur in the soil, the shear strength may reduce to the residual shear strength value.

The next and most recent full revision of BS 8002 is dated 2015 [78]. The revised text is based on limit state principles as set out in Eurocode 7; it provides designers with the guidance on how to implement the requirements of Eurocode 7. The code of practice also gives further details on how retaining structures should be designed according to limit state principles and provides information that is not included in Eurocode 7 but is of importance to practice in the U.K. Advances in retaining wall technology over the past 20 years are reflected in the fully revised text; particular guidance is given in relation to soil stiffness to enable serviceability limit states to be verified.

In British practice, guidance on design of earth retaining structures for deep excavations can also be obtained from publications by CIRIA, "Embedded retaining walls – guidance for economic design" C580 [79], and "*Guidance on embedded retaining wall design*" C760 [80], which supersedes C580. C760 provides a good practice guide on the selection and design of vertical embedded retaining walls to satisfy the requirements of Eurocodes. It covers temporary and permanent walls. Structural forms included are anchored, single and multi-propped retaining walls with embedment in competent soils and soft rocks.

3.10.2 Factor of safety approach

The approach to design of earth retaining structures in the 1950s was based on calculating the disturbing forces to be retained by the wall, namely the sum of the earth pressures on the back of the wall including the effects of any surcharging on the supported ground and the resisting forces that prevent excessive movement of the wall. CP2 [76] considered free-standing walls with three mechanisms, sliding that is resisted primarily by shear beneath the base of the wall, overturning by rotation about the front edge of the base and bearing capacity failure beneath the base. Geotechnical parameters were determined based on typical values. Mechanisms of movement were determined based on mobilization of the shear strength of the ground, and stability was achieved by adopting a factor of safety such that the ratio of restoring effects is greater than the disturbing effects by a factor of safety. Design of retaining walls for deep excavations generally includes resistance to movement of the wall by relying on an embedded part of the wall below the excavation level. The mechanisms of failure considered in CP2 are not readily applicable to deep earth-retaining structures. The factor of safety design approach concerns checking for stability and no overall failure. The governing geotechnical parameters are the strength of the soil. The approach does not consider movements, and the parameters governing the deformation of the ground were not required to be considered. BS 8002 (1994) [77] considers basement walls, excavation, support and retention systems appropriate to deep excavations. However, the latest version of BS 8002 [78] adopts the limit state method of design. The factor of safety method is no longer adopted for design of deep earth retaining structures and is replaced by the limit state approach.

3.10.3 Limit state approach

A limit state is a performance criterion to be adopted in design beyond which a structure or an element of a structure is assumed to become unfit for its purpose. There are two limiting states, serviceability and ultimate limit states. The approach was initiated for structural use of concrete in the U.K. in 1972 with the publication of CP110 [81], which concerned reinforced concrete structures and included partial factors for loads and materials. Limit state geotechnical design for earth retaining structures was introduced in the U.K. with the publication of BS 8002 in 1994 [77], but partial factors were not introduced for geotechnical design until later.

The ultimate limit state for a structure is deemed to have been reached when the structure, parts of it or the soil around the structure have yielded to form a failure mechanism in the ground or severe damage. Modes of failure to be checked in the design include sliding, overturning, bearing capacity, uplift and hydraulic failure. The ultimate limit state approach was analogous to the factor of safety approach. Generally, limiting earth pressures and factors of safety were adopted. Passive earth pressures (for calculating restraining movements), active pressures (for temporary works) and earth pressures (at rest for permanent works) were adopted for calculating applied ground pressures; factors of safety (carried over from the factor of safety approach) were also adopted.

The serviceability limit state of a structure is deemed to have been reached at the onset of excessive deformation or deterioration. Criteria for serviceability design include total and differential movements, cracking and vibration. The introduction of the serviceability limit state required calculations for earth retaining structures to include the use of deformation parameters for the ground and calculations for the movements and deformation of the structures and the ground. In order to take into account the different stiffness of retaining structures compared with the ground, estimates for movements and deformation require calculation of the interaction between the structures and the ground, now referred to as soil-structure interaction. For earth retaining structures for deep excavations, under working conditions, the retaining structure generally undergoes movement towards the excavation, and the movement develops progressively during successive stages of excavation. At the start of construction, the retained ground initially is at rest, and as excavation proceeds and the retaining structure moves forwards the lateral earth pressure drops; in some cases, it drops to the active earth pressure depending on the degree of movement. Below the level of the excavation, the embedded part of the retaining structure tends to move forwards. In principle, the lateral earth pressure increases to resist the movement depending on the degree of movement with a limiting passive earth pressure.

Calculations for soil-structure interaction by hand are very laborious and are suited to the use of computer programs. As described in Section 3.1, computer programs for soil-structure interaction came into being in the middle of the 1970s and provided the means by which serviceability limit state calculations could be carried out as a regular design procedure in 1994 when BS 8002 was introduced. In 1994, there were several different types of computer programs available, and generally they included simplifying assumptions. Methods in use at the time included associated stress and velocity fields methods, beam on elastic foundation method, boundary element method and finite element, or finite difference, method. Of these methods, the beam on elastic foundation method is still in use as a quick method, and the finite element and finite difference methods are popular. Commercial versions of the beam on elastic foundation method have been updated to include elastic and plastic behaviour of the ground and other improvements. The first in-house program that I used in 1976 was based on a beam not on an elastic foundation but on a medium that modelled both elastic and plastic behaviour.

For design of earth retaining structures for deep excavations when considering the serviceability limit state using unfactored geotechnical parameters and unfactored loading and typical of average values for parameters, the calculations provide an estimate of movements and deformations approximating those that may occur.

GCO of the HKSAR Government published "*Review of design methods for excavations*" in 1990 [75]. It provides further and detailed contemporary insight on the limit state approach, although some of its contents are now outdated and should be treated with caution. For example, several of the analytical methods described in Chapter 5 references are no longer in use. It is stated in Section 4.3 of the reference, "*finite element methods incorporating appropriate soil models … are presently too cumbersome and costly for general application, and accurate determination of the required soil parameters is difficult.*" The finite element method is now used generally for design of earth retaining structures. Importantly, at several places, it is assumed that on the passive side of a wall the soil is initially elastic until the passive pressure is reached. This assumption is valid at several metres' depth below excavation level, but it is misleading at shallow depths. The excavation process removes the overburden from above the current formation level and substantially reduces the vertical pressure for the first few metres of soil below the formation level. Within these few metres, the lateral earth pressure from the previous stage of excavation exceeds the passive pressure at the current stage of excavation and the passive soil pressure is developed, without any forwards movement being required while the previous stage of excavation took place [82]. This phenomenon is modelled by most of the commercial computer programs that are now in use.

3.10.4 Partial factor design

Limit State Design, incorporating partial factors, was introduced in the U.K. for design of concrete structures in CP110 [81]. Partial factors limit state design is recommended for geotechnical design in Eurocode 7 [29] and in BS8002 2015 [78]. The method is now in general use in many countries. The reason for recommending the use of partial factors is to reflect the variations in the parameter to which the partial factor applies. The type and properties of the ground vary considerably. For example, soft clay behaves quite differently to the behaviour of dense gravel. Ground investigation is a sampling procedure. It is quite common that samples from the same site vary in consistency and exhibit different strength and stiffness properties. The objective of the designer is to determine parameters for the ground that will reasonably predict the performance of the ground for purposes of serviceability and can be relied upon to resist failure in the case of ultimate limit state. For illustration, if a material parameter for design is chosen with 95% probability of not being exceeded, and if therefore a more onerous conditions would be 5% probability of occurrence and if the loading parameter was chosen on the same bases with 5% probability of being exceeded, the combined probability of more onerous conditions coinciding would be 5% of 5%, which is 0.25% or 1 in 400. When a material partial factor is applied to the material properties, thereby reducing the value for strength, and when a loading partial factor is applied to the loading, thereby increasing the loading, a limiting state can be considered. Then ensuring that the loading does not exceed the strength satisfies a check on the ultimate limit state.

Eurocode 7 [29] and BS 8002 2015 [78] make recommendations, but the designer should determine what parameters to use and how to use them. For example, in the case of soil strength, the designer can consider whether any very low results from tests are due to sample disturbance and can be disregarded or whether they indicate a weak layer within generally stronger soil. In the case of evaluating the strength of the ground, it is common for strength values to be plotted versus depth to determine a characteristic profile of strength versus depth and based on the plot to take a cautious estimate and to use it for the design. A cautious evaluation might be a profile with, say, only 5% of the values lying beneath the profile. When evaluating the parameters, in the case of the occurrence of unusually weak values, the designer can consider whether the unusually weak values are due to sampling disturbance and may be discarded, or due to weak lenses in the ground. If the occurrence of weak lenses or weak strata can be identified, then they may be specifically included as weak zones in the analysis and not affect the selection of the values of strength for the other zones up and down the wall. In the case of an earth-retaining structure with large wall panels, the loading is the sum of the pressure from the ground that is in contact all over the wall panel including weak and strong zones of the ground. In effect the total force on the wall panel could be estimated by taking the average of the values of the pressure calculated from the values for strength of the ground that is in contact with the wall and perhaps to adopt a profile of strength versus depth, which is moderately conservatively less than the average profile for the strength versus depth.

In principle, for the ultimate limit state, the adoption of partial factors can be calibrated to achieve the same result that would be obtained by adopting a global factor of safety. Taking a simplified example, by a partial material factor resulting in a reduced effect of 0.8 and a partial loading factor increasing by 1.2, the combined effect, or an overall factor of safety of 1.5 (=1.2/0.8). In the case of a design carried out in accordance with the factor of safety method, an earth retaining wall would be designed using unfactored parameters and the resulting effects such as bending moment in the wall would be increased by the factor of safety, say, by 1.5. If the geotechnical parameters are factored, as illustrated in Figure 3.4, not only is the intensity of loading increased but the length of loaded span is also increased as shown by the shaded area in Figure 3.4 compared to the shaded area for unfactored loads in Figure 3.3. The resulting bending moment with factored loads is close to the bending moment with unfactored loads times the factor of safety in this case. Because the factored loading extends further down the wall, the effect of a partial factor method is that the maximum bending moment is at a greater depth than when the factor of safety method is adopted. This condition is appropriate for deep earth retaining walls because in the event of the ground being weaker than was assumed in the design, the induced larger bending moment would have occurred at a lower location than was designed. As far as I know, the adoption of the partial factor method was a logical step and was not a remedy in the light of any earth-retaining structure for deep excavations having suffered distress or failure when designed properly.

Further insight and explanation of Eurocode 7 may be found in *"Decoding EuroCode 7"* by Bond and Harris [83]

3.11 Checking

"To err (make a mistake) is human, to forgive is divine" says an old proverb. In ancient Rome, architects, a name in Greek meaning bridge builders, were required to stand underneath the bridges that had been built under their responsibility while the temporary supports were taken away. This practice seems to have been intended to ensure that if the bridge builder had made a mistake the mistake would never happen again. In the modern civilised age, society does not expect engineers to stand under collapsing bridges, and society does not expect engineers to make mistakes. Therefore, it is only sensible that engineer's work should be checked.

3.11.1 Checking designs

The simplest level of checking is a regular review by a supervisor or team leader of work done at a given stage. However, since mistakes can be made, there should be a thorough detailed check. An independent checker should take the calculations and run a mathematical check. However, increasing use of computer programs and spreadsheets leads to complexity and multiplicity of calculations, and a manual check is often impractical except for design calculations for minor structures. Therefore, calculations are more usually checked in detail by an independent engineer who checks the selection of parameters and the input data and runs parallel calculations, perhaps with different software. It is fundamental to checking that computer programs and spreadsheets be verified for accuracy and that designers be well versed in how to use the software properly. Another level is checking by an independent checker who takes drawings and conducts its own independent set of design calculations. Such checks are for accuracy and compliance with requirements such as standards, codes of practices and project design requirements. It is usually a requirement that a designer produce a design report setting out the basis of design, the design requirements that have been adopted and parameters used for the design. It is also usual for an independent checker to produce a similar report highlighting any deficiencies the designer should amend; then the checker should sign off that the deficiencies have been rectified. A thorough independent check based on design outline drawings involves a considerable amount of work of the order of 60% of the work required of the designer who in addition to design calculations is required to carry out the planning and design development. Checkers could be deployed in-house, but their independence might be criticised; independent checkers from a separate company are sometimes required by the employer.

For all projects, it is important to maintain Quality Control (QC). In simple terms, QC keeps paperwork in order. For purposes of design of deep excavations, QC mostly relates to keeping up to date with any changes in the employer's requirements for the design and interfaces with related designers such as architects, structural engineers and others who are concerned with the permanent structure that is built in the deep excavation.

Hong Kong and Singapore have regulatory control of design and checking. Independent Checking Engineers (ICE) are separately appointed and required for public works. For private building works in Hong Kong, designs are submitted in detail to the BA and are reviewed by BD. The BD employs structural engineers and surveyors who review the submitted plans; when they are accepted, approval is given on behalf of BA. GEO provides an advisory service to BD. BD and GEO do not conduct detailed checking, and it is the responsibility of RSEs and RGEs to check the adequacy of their own work. Regulatory control of building works in Hong Kong is discussed in Section 3.7. In Singapore, BCA has established list of Accredited Checkers (AC) whose role is to check and certify plans to be submitted to BCA for approval before permission is granted for construction. In Singapore, both public and private works are subject to checking by an AC. The statutory duty of ICEs or ACs is specifically to ensure compliance with standards, codes of practice and regulations with respect to public safety.

3.11.2 Peer reviews

It is good practice to ensure that quality of a design is reviewed. The design team leader should review the design as it develops. For large, complicated or unusual projects, it can be beneficial to conduct reviews by specialists. The term "Peer Review" is often used for such overall reviews. The word "Peer" implies equality with the project leader and peer reviewers should be well experienced in their field of practice and can be invited in-house or externally. Important stages of a project for conducting peer reviews are as follows:

i. When bidding for appointment as Engineer in the hope of winning the commission,
ii. Early in the design process when the concept of the project is about to be selected,
iii. Before the first deliverable to the employer to ensure that the concept is adopted as intended and
iv. When drafting the Contract Documents to ensure that the project and complexities are properly defined

Peer reviews usually start with a briefing for the reviewers. Reviewers must be informed of the project brief, site information and setting. There follow wide-ranging evaluation options and interaction with the design team. Stages i and ii tend to consider concepts. Stage iii tends to focus on the development of the chosen option, what details are adopted and how risks have been evaluated, as well as the approaches taken for mitigation of risks. Stage iv, reviewing draft contract documents, is often done by specialists individually reviewing relevant sections of the documents.

Peer reviews provide the opportunity for lateral thinking. There are many completed projects, and pressure to design quickly leads designers to adopt previously used concepts. One design does not suit all purposes for all sites. For example, for a shaft for recovery of a TBM to be completed with a length of cut and cover tunnel built within temporary retaining walls, a designer might choose the tunnel to be independent of the ELS because the roof and floor spans of the tunnel would be shorter than using the temporary walls as permanent walls, and permanent design requiring additional steel reinforcement for crack control would be limited to the tunnel permanent structure and not include the more extensive temporary retaining walls. The construction specialist in this case preferred to use diaphragm walls as temporary and as permanent walls and have wider roof and floor slabs because that solution would be faster to construct.

3.11.3 Legal liability

Although the designer is responsible overall for the design and would be liable for any deficiency, in the event that the design is found to be deficient, the independent checker is likely to be charged as responsible for a proportion of the liability, for having accepted and certified a deficient design.

3.12 Programming activities for construction

For all projects, the engineer is usually required to produce an estimate of the cost and a programme for construction. For large projects a fully Costed Implementation Programme (CIP) is required. A CIP that is prepared by an engineer should be similar to the programme and costing that the tenderers need in order to prepare their tender bid. A planning exercise is carried out for the construction. Plans are drawn up for utilisation of the site: entrance sand exits, site offices, materials stores, equipment stores, workshops, slurry treatment plant if required and any other immovable plant and so on. Operations of moveable plant are laid out including working space and haulage routes. The number of each type of plant and plant utilisation is programmed allowing for mobilisation, downtime, repairs. Production rates are worked out based on hours per day and days per week of working time with an allowance for bad weather and including what work can be carried out at night time and any noise restriction time and any other restrictions such as night time working for occupation of roads or railways for relocating utilities or for installing protective measures. Plant related costs are built up for purchase and resale or hire of plant, for fuel and operating costs, replacement of parts and servicing. Costs are estimated for supply of materials and for disposal of waste, including any licences required, for example for disposal of bentonite and contaminate waste. The overall programme includes lead time to procure or mobilise plant and materials and costs for quality control, testing, site supervision and contractor's overheads.

For large sites, the work might be divided into two or more contracts. Then there is a need for a large site management team and supervisory staff. With many activities taking place on site there are interfaces to control, with permits for given areas, and sometimes a need for traffic management, not only around the site but also within the site itself.

References

[1] Peck, R.B. (1969) Deep Excavations and Tunnelling in Soft Ground. *Proc. of Seventh Int. Conf. on Soil Mechanics and Foundation Engineering*, Mexico City, State of the art vol., pp. 225–290.

[2] GCO. (1990) *Review of Design Methods for Excavations*, GCO Publication. No. 1/90. Government of HKSAR, Hong Kong.

[3] Pappin, J., Endicott, L.J., & Clarke, J. (2006) Deep Excavation in Hong Kong – Design and Construction Control, H.K.I.E. Seminar, Hong Kong, January. 2006. http://hkieged.org/HKIE/download/JP%20JE%20and%20JC.pdf

[4] GCO. (2017) *Guide to Retaining Wall Design (Geoguide 1)* (Continuously Updated E-Version released on 29 August 2017). Geotechnical Engineering Office, Civil Engineering and Development Department, HKSAR Government, 245 p. www.cedd.gov.hk/filemanager/eng/content_106/eg1_20170829.pdf

[5] Winkler, E. (1867) *Die Lehre von der Elasticitaet und Festigkeit*, H. Dominicus, Prague, 1 January 1867.

[6] Kurrer, K.E. (2009) *The History of the Theory of Structures: From Arch Analysis to Computational Mechanics*, Ernst & Sohn, Berlin. ISNB 978-3-433-01838-5.

[7] WALLAP. Commercial Software. www.geoengineer.org/software/wallap

[8] Zienkiewicz, O.C. & Taylor, R.L. (1989) *The Finite Element Method*, 4th Edition. McGraw-Hill, London. ISBN 0-07-099193-6.

[9] Simpson, B. (1973) Finite Elements Applied to Problems of Plane Strain Deformation of Soils, PhD Thesis, University of Cambridge.

[10] OASYS. Commercial Software. www.oasys-software.com/

[11] ITASCA. (1987) *User's Manual for FLAC – Fast Lagrangian Analysis of Continua*, ITASCA Consulting Group Inc., USA. www.itascacg.com/software/flac

[12] Endicott, L.J. (2017) 7th Lumb Lecture 10th October 2012 Peter Lumb's Legacy, Soil Mechanics = Simple Concepts + Mathematical Processes + Lateral Thinking. *Engineering Journal of the SEAGS & AGSSEA*, Volume 47 Issue 3, September 2016, pp. 37–50. ISSN 0046-5828.

[13] COI Report. (2005) *Report of the Committee of Inquiry into the Incident at the MRT Circle Line Worksite that Led to the Collapse of the Nicoll Highway on 20 April 2004*, Ministry of Manpower, Singapore. (Call no.: RSING 363.119624171 SIN).

[14] Burland, J.B. (2012) Chapter 26 Building Response to Ground Movements. *I.C.E. Manual of Geotechnical Engineering*, Volume 1, pp. 281–295. ISBN 978027757081.

[15] Burland, J.B., Standing, J.R., & Jardine, F.M. (2001) *Building Response to Tunnelling – Case Studies from the Jubilee Line Extension*, London (SP200), Thomas Telford, London, July 2001. ISBN: 978-0-7277-3017-6.

[16] Endicott, L.J. (2013) Spatial Variations in Groundwater Response during Deep Tunnelling. *Proceedings of 18th SEAGS and Inaugural AGSSEA Conference*, Leung, C.F., Goh, S.H. & Shen, R.F. (Eds.). Singapore, May, p. 359.

[17] Morton, K., Leonard, M.S.M., & Cater, R.W. (1980) Building Settlements and Ground Movements Associated with Construction of Two Stations of the Modified Initial System of the Mass Transit Railway, Hong Kong. *Proceedings of 2nd International Conference on Ground Movements and Structures*, Cardiff, pp. 788–802.

[18] PNAP-22 Dewatering in Foundations and Basement Excavation Works. (2009) BA HKSAR Government, August. www.bd.gov.hk/doc/en/resources/codes-and-references/practice-notes-and-circular-letters/pnap/APP/APP022.pdf

[19] Legislative Council Secretariat IN26/02-03. www.legco.gov.hk/yr02-03/english/sec/library/0203in26e.pdf

[20] SPU SP199. (2001) Response of Buildings to Excavation-induced Ground Movements. *Proceedings of the International Conference "Response of Buildings to Excavation-Induced Ground Movements,"* Imperial College, London, July.

[21] Cheuk, J.C.Y., Lai, A.W.L., Cheung, C.K.W., Man, V.K.M., & So, A.K.O. (2013) The Use of Jet Grouting to Enhance Stability of Bermed Excavation. *Proceedings of 18th International Conference on Soil Mechanics and Geotechnical Engineering,* Paris, pp. 1255–1258.

[22] Jack Crooks: Principal, Golder Associates, Calgary, Canada.

[23] Endicott, L.J. (2006) Nicoll Highway Lessons Learned. *Proceedings of International Conference on Deep Excavations,* Singapore, June.

[24] Peck, R.B. (1969) Advantages and Limitations of the Observational Method in Applied Soil Mechanics. *Geotechnique,* Volume 19 Issue 2, pp. 171–187.

[25] Spross, J. & Johansson, F. (2017, May) When Is the Observational Method in Geotechnical Engineering Favourable? *Structural Safety,* Volume 66, pp. 17–26. Elsevier. doi:10.1016/j.strusafe.2017.01.006.

[26] Endicott, L.J. (1980) Aspects of Design of Underground Railway Stations to Suit Local Soil Conditions in Hong Kong. *Hong Kong Engineer,* Volume 8, pp. 29–38.

[27] Geotechnical Information Unit. Geotechnical Engineering Office, Civil Engineering Department, HKSAR Government. www.cedd.gov.hk/eng/public-services-forms/geotechnical/library/index.html

[28] *Geotechnical Manual for Slopes.* (1984) 2nd Edition. Geotechnical Engineering Office, Civil Engineering Development Department. HKSAR Government. www.cedd.gov.hk/eng/publications/geo/geo-gms/index.html

[29] *Eurocode 7: Geotechnical Design.* (2013) European Union. ISBN 978-92-79-33759-8.

[30] Essex, R.J. (2007) *Geotechnical Baseline Reports for Construction,* American Society of Civil Engineers, Reston, VA, ISBN13:978-0-7844-0939-5. https://ascelibrary.org/doi/pdf/10.1061/9780784409305.fm

[31] SCMP. (2001) Lawmakers Slam $750m Payout, *South China Morning Post,* 5 October 2001, p. 1.

[32] SSDS Stage 1 Breaking through all obstacles Hong Kong Engineer. www.hkengineer.org.hk/program/home/articlelist.php?cat=cover&volid=6tp

[33] SCMP. (2001) Drainage. Completion of the Harbour Area Treatment Scheme Stage 1. *South China Morning Post,* 10 December 2001, p. 1.

[34] Endicott, L.J. (2010) Managing Geotechnical Risk in Deep Tunnels in Hong Kong. *Proc. 11th Congress of the I.A.E.G.,* New Zealand, September, p. 391.

[35] Morwood, R., Scott, D., & Pitcher, I. (2008) Alliancing a Participant's Guide. Maunsell-AECOM, Brisbane, Australia. ISBN 978-0-646-50284.

[36] *Building Ordinance Cap. 123,* HKSAR Government. www.elegislation.gov.hk/hk/cap123

[37] *Buildings (Administration) Regulations Cap. 123A,* HKSAR Government. www.elegislation.gov.hk/hk/cap123A

[38] Practice Notes for Authorized Persons, Registered Structural Engineers and Registered Geotechnical Engineers. (2019) BD, HKSAR Government. www.bd.gov.hk/en/resources/codes-and-references/practice-notes-and-circular-letters/index_pnap.html

[39] *Buildings (Construction) Regulations Cap 123B*, HKSAR Government. www.elegislation.gov.hk/hk/cap123B?xpid=ID_1438402645257_002

[40] List of British Standards. (2019) British Standards Institution. www.bsigroup.com/en-HK/Standards/british-standards-online/

[41] List of Eurocodes and National Annexes in the U.K. www.eurocoded.com/mod/page/view.php?id=18

[42] *BS 1377: Part 2: 1990 Methods of Test for Soils for Civil Engineering Purposes – Classification Tests*, British Standards Institution.

[43] GEO publications. www.cedd.gov.hk/eng/publications/geo/index.html

[44] PNAP APP-57 Requirements for an Excavation and Lateral Support Plan. (2012) BD, HKSAR Government.

[45] PNAP APP-30 Development in the Mid-Levels Scheduled Area. (2005, December) BD, HKSA Government.

[46] PNAP APP-61 Development in Areas Numbers 2&4 of the Scheduled Areas. (2009) BA, HKSAR Government, Aug.

[47] PNAP APP-24 Railway Protection Railway Protection Ordinance. (2013, May) BA. HKSAR Government.

[48] PNAP APP-62 Protection of Sewage and Drainage Tunnels. (2018, March) BA. HKSAR Government.

[49] PNAP APP-28 Requirements for Qualified Supervision of Site Formation Works, Excavation Works, Foundation Works on Sloping Ground, and Ground Investigation Works in Scheduled Areas. (1997, December) BA, HKSAR Government.

[50] PNAP APP-48 Requirements for Qualified Supervision of Structural Works. (2015, December) Foundation Works and Excavation Works, BD, HKSAR Government.

[51] Plant, G.W., Covil, C.S., & Hughes, R.A. (Eds.). (1998) *Site Preparation for the New Hong Kong International Airport*, Thomas Telford, London, pp. 25, 26, ISBN0-7277-2696-X.

[52] Civil Engineering Library, CEDD, HKSAR Government. https://www.cedd.gov.hk/eng/public-services-forms/geotechnical/library/index.html

[53] *Building Control Regulations*. (2019) Part V 42(f), BCA, Singapore. https://sso.agc.gov.sg/SL/BCA1989-S666-2003#pr42-

[54] Geographic Information System. www.esri.com/en-us/what-is-gis/overview

[55] BIM https://en.wikipedia.org/wiki/Building_information_modeling

[56] Endicott, L.J. & Tattersall, J.W. (2009) The Use of Geological Models and Construction Data to Estimate Tunnelling Performance with respect to Reducing Inflow of Ground Water, *Proceedings of Hong Kong Tunnelling Conference*, November.

[57] Groves, P., Cascante, G., Dundas, D., & Chatterji, P.K. (2011) Use of Geophysical Methods for Soil Profile Evaluation. *Canadian Geotechnical Journal*, Volume 48 Issue 9, pp. 1364–1377. doi:10.1139/t11-044.

[58] *From Geophysical Parameters to Soil Characteristics*. (2009) FP7-DIGISOIL Project Deliverable D2.1 No. FP7-DIGISOIL-D2.1. BRGM. March.

[59] Standard Practice for Environmental Site Assessments: Phase I Environmental Site Assessment Process. ASTM E1527-13. www.astm.org/Standards/E1527.htm

[60] ASTM. (2019) Standard Practice for Environmental Site Assessments: Phase II Environmental Site Assessment Process. ASTM E1903-11. www.astm.org/Standards/E1527.htm

[61] *PLAXIS 2D User's Manual*, Bentley Systems, Incorporated. www.plaxis.com/support/manuals/plaxis-2d-manuals/

[62] Chan, G., Lui, J., Lam, K., Yin, K., Law, L.R., Chan, A., & Hasle, R. (2004) Shaft Grouted Friction Barrette Piles for a Super High-rise Building. *The Structural Engineer*, Volume 82 Issue 2. www.istructe.org/journal/volumes/volume-82-(published-in-2004)/issue-20/shaft-grouted-friction-barrette-piles-for-a-super/

[63] *Uniform Building Code, Vol. 2: Structural Engineering Design Provisions International Code Council, U.S.A.* ISBN-13: 978-1884590894, ISBN-10: 1884590896.

[64] *Guidelines for Determining Design Basis Ground Motions*, Electric Power Research Institute, Paolo Alto, CA, Vol. 1, pp. 8–1 through 8–69.

[65] Richter, C.F. (1958) *Elementary Seismology*, W.H. Freeman, San Francisco, CA.

[66] GEO Publication No. 1/2012 Review of Earthquake Data for the Hong Kong Region. (2012) Geotechnical Engineering Office, Civil Engineering Department, HKSAR Government.

[67] Ishiyama, Y. (2012) Introduction to Earthquake Engineering and Seismic Codes in the World, July. http://iisee.kenken.go.jp/

[68] *Highway Bridge Design Specification Part V Seismic Design.* (2002) Japan Road Association.

[69] Loma Prieta Earthquake. https://en.wikipedia.org/wiki/1989_Loma_Prieta_earthquake

[70] Earthquake History. www.earthquakesafety.com/earthquake-history.html

[71] NCHRP Report 61. (2008) *Seismic Analysis and Design of Retaining Walls, Buried Structures, Slopes, and Embankments 2008 NCHRP Report 61*, The National Academies Press. ISSN 0077-5614, ISBN: 978-0-309-11765-4 Transport Research Board (c) USA.

[72] Idrisss, I.M. & Boulanger, R.W. (2014) *CPT and SPT Based Liquefaction Triggering Procedures*. Report No UCD/CGM-14/01, U.C. Davis. https://faculty.engineering.ucdavis.edu/boulanger/wp-content/uploads/sites/71/2014/09/Boulanger_Idriss_CPT_and_SPT_Liq_triggering_CGM-14-01_20141.pdf

[73] Hitchings, W. (1979) *West Gate*, Outback Press, Melbourne, Australia. ISBN 9780868882260.

[74] Endicott, L.J. (2013) Examination of Where Things Went Wrong: Nicholl Highway Collapse, Singapore. *7th International Conference on Case Histories in Geotechnical Engineering*, Wheeling, IL, April 29–May 4.

[75] *Review of Design Methods for Excavations*, GCO Publication No. 1/90. Geotechnical Engineering Office, Civil Engineering Development Department, HKSAR Government.

[76] The Institution of Structural Engineers. (1951) *The Civil Engineering Code of Practice No.2 (CP2)*, The Institution of Structural Engineers, London.

[77] BS 8002. (1994) *Code of Practice for Earth Retaining Structures*, British Standards Institute, UK. ISBN: 0 580 22826 6.

[78] BS 8002. (2015) *Code of Practice for Earth Retaining Structures*, British Standards Institute, UK. ISBN: 978 0 580 86678 4.

[79] C580. (2003) *Embedded Retaining Walls – Guidance for Economic Design*, CIRIA, London, UK. ISBN: 0 86017 580.

[80] C760. (2015) *Guidance on Embedded Retaining Wall Design*, CIRIA, London, UK. ISBN: 978-0-86017-764-7.

[81] CP110-1. (1972) *Code of Practice for the Structural Use of Concrete. Design, Materials and Workmanship*, British Standards Institute, UK. ISBM 0 580 07488 9.

[82] Endicott, L.J. & Cheung, C.T. (1991) Temporary Earth Support. *Proc. of Seminar on Lateral Ground Support Systems*, H.K.I.E., May 1991, pp. 39–49.

[83] Bond, A. & Harris, A. (2008) *Decoding EuroCode 7*. Taylor & Francis. ISBN 978-0-415-40948-3.

Chapter 4

Contracts and construction

In order for a deep excavation to be carried out, a construction company needs to be engaged by contract. A contract is drawn up and assigned to a construction company who then becomes the contractor. Usually, large civil engineering projects such as deep excavations require supervision and a supervising engineer may be appointed. If a contractor is required to self-supervise then the employer will probably retain a general consultant inspect the work, certify work completed for payments and provide technical advice to the employer.

4.1 Contracts

In order for an excavation to begin a contract has to be let. A contract sets out the duties and obligations of The Employer, then duties and obligations of the contractor and the role and authority of The Engineer or The Supervisor. The use of capitals refers to the role specifically for a given contract. In this book lower case refers to the general case. This section is aimed at introducing the concepts and principles of contracts for an engineer engaged in deep excavations and is not intended to be comprehensive or authoritative. For insight the reader is recommended to further reading and/or specialist advice.

4.1.1 Forms of contract

There are different forms of contract that could be adopted for building work. For deep excavations it is appropriate to use a civil engineering form of contract, which addresses the inherent variability of the ground in addition to other risk items such as variations in costs of fuel and materials. For large projects, such as deep excavations usually the employer has a preferred form of contract. For example, governments have standard forms of contract that they require. Internationally, there are several forms of contract such as the Joint Contracts Tribunal form, the FIDIC form published by the International Federation of Consulting Engineers and the I.C.E. Form, published by the Institution of Civil Engineers (I.C.E.), which is now largely replaced by the New Engineering Contract (NEC). NEC form is in use in U.K., Australia, New Zealand, South Africa and Hong Kong. A contractor may let sub-contracts to engage sub-contractors for specific parts of the works. A sub-contractor contract should closely follow the contract on a back-to-back basis; otherwise, difficulties can arise, for example, when additional works are required.

4.1.2 Extent of contract

For deep excavations, especially in urban areas, the scope of a contract usually includes not only excavation but also building permanent works within the excavation. For single uncomplicated underground structures, one contractor may be appointed for all of the work. For larger underground projects, a contractor may sublet substantial parts of the works to one or more sub-contractors. In South East Asia, quite often, sub-contractors sublet to sub-sub-contractors who in turn sublet. The most complicated subletting that I encountered was nine levels of sub-contracting. The ninth level was a gang of workmen who in effect performed piece-work getting paid a sum for each task completed and were not employed as hired labour. For large complicated projects, a management contract may be adopted whereby the management contractor lets and manages contracts for packages of the work. The West Kowloon Station in Hong Kong [1], with thousands of people on site, had two main contractors with 24 other contractors and many sub-contractors operating on a site of about 12 hectares. Not least amongst the tasks during construction was site management with so many workers and so many items of plant operating.

Contracts for deep excavations are sometimes based on an engineer's design, which a contractor builds according to details produced on behalf of the Employer by a designer. Some contracts for deep excavations are based on a scope to both design and build whereby tender documents generally include a reference design for purposes of expanding on an employer's requirements. Tenderers are invited to provide their own design, which may be based on a reference design or may adopt a different design for the same purpose.

There are several basic forms of contract regarding payment to a contractor including the following:

- Lump sum, fixed price contract.
- Measurement contract.
- Cost-reimbursement contract.
- Alliance contract

A lump sum contract means that a contractor only gets paid the lump sum agreed in the contract, irrespective of the quantity of work that is done. Interim payments are based on completed work in accordance with a Bill of Quantities. Work is in accordance with specifications and supplied drawings. This form of contract is suited to projects where the scope and nature of the work is well defined. At the outset, when preparing a tender, the tenderer knows precisely what is required. A lump sum priced contract can be used but with some agreed rates such that measurement can allow for variations that sometimes occur.

Measurement contracts include a schedule of rates with approximate quantities, and payment is made according to the amount of work that is done. This form of contract has flexibility when the exact quantities of work to be carried out are not know at the outset. Sometimes this form of contract is adopted even when the quantities of work can be reliably estimated at the award of contract.

Cost-reimbursement contract is when the Employer agrees to pay the Contractor for the direct cost and reimbursable expenses. It is used when the scope of work, or the costs for performing the work are well known.

Alliance contracts, also called partnering contracts, are intended to share risks and motivation between contracting parties. Alliance contracts have been adopted in Australia for several years [2]. In the wider context, the parties include employers, designers, contractors and sub-contractors, supervisors and suppliers since they all have parts to play in managing the risks and determining the costs and profitability and time for completion to a greater or lesser degree. Of contracts for deep excavations often I have heard it said that Government owns the land, a contractor has no interest in the type of ground other than getting it excavated as soon as possible, why should he accept any geological risk on land that is the property of Government? Alliancing and partnering aim to get a good deal for everybody.

The primary purposes of a contract are to identify what work is to be carried out by a contractor, what he will be paid for and how he will be paid for work done. What work the contractor is required to do, inter alia, includes forming the excavation, installing temporary and/or permanent ELS, monitoring, diverting utilities, mitigation measures and so on. All of the requirements of what is to be done by the contractor have to be clearly defined on drawings and in specifications that are included in the contract. It is also necessary for the contract to state the contract period, the time for completion and dates for any intermediate milestones to be met. In return, the employer contracts to pay for work that is done. How the payments are to be made is laid out in a method of measurement and a schedule for payments. For civil engineering work it is usual now to adopt some form of measurement with cost items or a schedule of rates. For large projects for HKSAR Government, the rates are often adjusted in accordance with basic costs for fuel, labour and materials linked to published indices. By this means the commercial risk on prime costs is not built into a contractor's bid price, and these risks are assumed by the employer on the expectation of getting a more competitive tender and, in effect, to pay for what he gets.

4.1.3 Contractor's all geotechnical risks contract

Dean Lewis writes "*Under common law the risk of unforeseen ground conditions rests exclusively with a contractor*" [3]. Therefore in any contract that does not state otherwise, common law applies and the contractor takes all the geological risk. For works in the ground there is an inherent natural risk in the ground itself. It seems somewhat incongruous that for contracts let with price adjustment for basic costs as mentioned above the Contractor was required to assume all geological risk, which has potentially far greater impact. For example, in the HKSAR Government's "*General Conditions of Contract for Civil Engineering Works*" [4], Clause 13 (1) states

> The Contractor shall be deemed to have examined and inspected the site ... the nature of the ground and sub-soil ..., and Clause 13 (2) states No claim by the Contractor for additional payment shall be allowed on the ground of any misunderstanding in respect of the matters referred to in sub-clause (1).

Whereas this might be a reasonable requirement for tenderers to evaluate the surface conditions of the site that could be seen during a site visit, the requirement imposed a heavy burden on tenderers if the Employer provided very little information about the sub-surface conditions and especially if the Employer provided no information about the sub-surface conditions at all. In order to identify sub-surface ground conditions, for some civil engineering projects in the past, tenderers carried out a few bore holes at their own expense. Sometimes the tenderers clubbed together to pay for some bore holes, but what could be done in a short period at the time for tender was often not a great deal.

In my own experience for the excavation for an underpass at Birmingham U.K., a tenderer wanted more information than The Employer provided and conducted more some bore holes and conducted some additional strength tests to be confident to offer an alternative tender adopting an open cut excavation with steep temporary slopes in relatively firm ground of Keuper Marl whereas the conforming design for temporary works was to provide an ELS system with bored piles and lagging. The tenderer's gamble paid off. His investment in some focused investigation resulted in confidence to offer a winning low bid price with a savings to the Employer and, no doubt, a suitable margin of profit for himself. On other projects, it was amazing that some employers paid for ground investigations for a designer to use but would not disclose the ground investigation reports to tenderers under the mistaken apprehension that the eventual contractor might use the bore hole results provided by The Employer in claims against The Employer. I recall that the British Courts of Justice has taken a dim view on employers withholding site information in this way, and it is more forthright to supply all the ground investigation reports that are available with a reminder that ground conditions may vary with time and can be expected to vary between bore hole locations. Indeed, in some locations the ground conditions could change from those encountered by the investigations carried out before the award of the contract. For example, in recent land reclamation when consolidation and settlements are on-going it is to be expected that the condition of the ground will change with time as on-going consolidation takes place. Excess pore ground water pressure should reduce, the surface of the ground should settle and the consolidating soils should gain in strength.

4.1.4 Towards equitable sharing of risks

In Hong Kong in the 1990s, the MTRCL took a more equitable approach in their conditions of contract at the time and would consider legitimate claims for unforeseen ground conditions and at Clause 38 [3] there is a definition of foreseeability stating

> If however during the Execution of the Works the Contractor shall encounter within the Site physical conditions (Other than weather conditions or conditions due to weather conditions) or artificial obstructions which conditions or obstructions he considers could not reasonably have been foreseen by an experienced contractor at the date of the Letter of Clarification.

In Hong Kong, MTRCL gained a reputation of addressing claims quickly and with a sense of fairness.

For many civil engineering projects such as deep excavations and for tunnels, the geological risk in the 1990s was, and remains, high. The cost, equipment to be used and time required to excavate deep excavations and tunnels depends on the type of ground and ground water conditions encountered. In the past, with little information about the ground condition on which to base a tender and with competitive tendering based entirely or largely on price, a high price with included pricing for geological risk was unlikely to be low enough to be awarded the contract and a low priced contract with little or no allowance for geological risk exposed a contractor to the possibility of a loss if ground conditions turned out to be worse than what he had priced for. Despite the provisions of HKSAR Government's Clause 13 putting all the geological risk on a contractor, there were many substantial claims by contractors relating to ground conditions on HKSAR

Government's contracts and cost of litigation were commensurately high. Various professional bodies including insurers and civil engineers realised the folly of continuing to place all, or a large proportion, of the geological risk on a contractor while awarding a contract on the lowest tendered sum, or even the second lowest tendered sum in infrequent instances. In the U.S.A., the American Society of Civil Engineers (ASCE) sought to formulate a structure for achieving a more equitable sharing of risk whilst achieving value for money by competitive bidding [6]. ASCE proposed the inclusion in contracts of Geotechnical Baselines. In the 1970s it became the practice in the U.S.A. to incorporate Geotechnical Interpretative Reports (GIRs) in contract documents; however, the litigious pursuit of claims in U.S.A. on civil engineering projects did not improve. In 1997 ASCE and the Construction Institute published a book setting out suggested guidelines for Geotechnical Baseline Reports (GBRs) for construction, (ASCE Guidelines). The preface of the book stated that *"Poorly written and ambiguous interpretative geotechnical reports, and inconsistencies between the interpretative report and other Contract Documents, were doing more harm than good in the effort to avoid and resolve construction disputes"* The Recommendation of The ASCE Guidelines is stated as

It is recommended that a single interpretative report be included in the Contract Documents and be called a Geotechnical Baseline Report (GBR). The primary purpose of the GBR is to establish a single source document where contractual statements describe the geotechnical conditions anticipated (or to be assumed) to be encountered during underground and subsurface construction. The contractual statement(s) are referred to as baselines. Risks associated with conditions consistent with or less adverse that the baselines are allocated to the Contractor, and those materially more adverse than the baselines are accepted by the Owner. Other important objectives of the GBR are to discuss the geotechnical and site conditions related to the anticipated means and methods of constructing the underground elements of the project.

GBRs have been in use in the U.S.A. increasingly since 1997 with some measure of success. GBRs were introduced to Hong Kong by the HKSAR Government in about 2006 and implemented in 2008 by the Drainage Department [7]. GBRs were introduced to Singapore at about the same time. In Singapore the name Geotechnical Interpretative and Baseline Report (GIBR) has been adopted but as set out in the ASCE Guidelines, referenced above, a GBR comprises " ... *single interpretative report* ... " There is no material difference in content and scope between a GIBR as adopted by Land Transport Authority (LTA) for the Mass Rapid Transit (MRT) in Singapore than the GBRs adopted by MTRCL in Hong Kong for the same purpose. Some contracts including GBRs and GIBRs have been concluded in Hong Kong and in

Singapore. I am aware that there have been some arbitrations on contracts with GBRs; however, the hearings were in camera. In my experience GBRs and GIBRs have served to define some, if not all, of the geotechnical baselines for the contracts; the introduction of a new document and a new practice is not without some pitfalls, such as poor drafting and misinterpretation. Overall my impression is that the adoption of GBRs and GIBRs has facilitated the acceptance of legitimate claims.

4.1.5 New forms of contract

In response to many claims on changed ground conditions internationally, the I.C.E. in the U.K. drafted a New Engineering Contract (NEC). The first NEC was published in 1993 [8]. It was a radical departure from previous building and engineering contracts, being written in plain language and designed to stimulate rather than frustrate good management. The NEC sets out a framework whereby contracts legally define the responsibilities and duties of the Employer and of the Contractor. The NEC system offers a suite of options that may be adopted. The HKSAR Government considered adopting NEC3 in about 2008 and after a successful trial for construction of the Tin Shui Wai Hospital announced in 2016 that all contracts shall follow NEC3 as far as is practicable [9].

Fundamental to a contract is the time for completion. There are usually provisions for late completion that include liquidated damages. These are charges by an employer for late completion, and they are usually specified in dollars, sometimes millions of dollars, per day in the contract.

For major projects, tendering requires pre-qualification or short-listing of the tenderers to demonstrate that they have adequate resources and prior experience in work of the same nature. The contract is usually awarded to the lowest bidder. Prior to the award of the contract an engineer usually evaluates the tenders in terms of compliance with the conditions of tender, adequacy of resources, methodology and tendered programme. The award of contract is based on price, but sometimes there is a formula whereby technical marks are given to the tenderers based on evaluation of their technical proposals as well as the price.

4.2 Construction

Construction of deep basements requires many skills. The main player is the Contractor. He takes possession of the site and takes responsibility for the site until he hands it back to the employer. The Contractor may engage sub-contractors whom the Contractor shall manage and take overall responsibility. The construction work is supervised, generally by Resident Site Staff (RSS) who are employed either by the Employer or by the Engineer, the specific named person/s under the contract.

4.2.1 Manuals

As described in Section 3.7 in Hong Kong, BA exercises control over building works primarily for purposes of public safety. As a consequence, regulations and practice notes are focused only on issues affecting, directly or indirectly, public safety. In addition to concerns about public safety, HKSAR Government is concerned with management of the project with respect to good practices, cost control, timely completion and avoidance of nuisance. Management Handbooks [10] produced by Development Bureau (Works Branch) of HKSAR Government give comprehensive directions on procedures for construction [11]. Such management is good practice and applies generally to civil engineering projects. This section is based on and summarises construction procedures and management for public works in Hong Kong.

4.2.2 Insurance and bonds

After the award of contract, there should be insurance in place. Sometimes the Contractor, and sometime the Employer provides the insurance for the project. The Contractor is required to provide a performance bond, which is often serviced by a bank and is in place to assure the Employer of performance by the Contractor [12]. During the currency of a contract the Employer may exercise retention moneys thereby increasing the monetary value of the security. If the Contractor is in default the Employer may use the performance bond to discharge damages sustained by the Employer arising from the default of the Contractor. Upon satisfactory completion of the project bonds and retention monies are released to the Contractor.

4.2.3 Site security and safety

For guidance on site safety, Works Bureau of HKSAR Government has published a Site Safety Manual [13]. Within a stipulated period of time after the award of contract the site is handed over to the Contractor, sometimes by stages, from which time the Contractor becomes responsible and occupier of the site. He is responsible for security of the site and the consequences of his subsequent operations until hand over of the site after completion of construction. Safety is always very important, and amongst the first tasks is to fence the site and to post security personnel to make the site secure. In urban areas a fence may not be sufficient and hoarding, comprising a temporary board fence, may be erected round the site. Notices are exhibited giving contact names and numbers in case of emergency. Measures for safety include a safety officer, registration of persons entering and leaving the site and on congested sites, registration of persons within respective areas of the site. Safety must be a culture, stemming from the management and sustained at working level. Many of the site labourers have few skills and become complacent about their personal safety. At changes of shift, briefings of the workforce called "tool box meetings" are held for the purpose of advising crews and works supervisors of their daily tasks and about what else is taking place on the site during the day and to remind them about safety precautions and procedure. Everybody on site should wear personal protective gear including a hard hat, safety footwear and other items depending on the nature of the work that is going on. A lot of excavation is done in the open. When structures are built in the excavation, enclosed spaces are sometimes created and these require precautions to ensure that there is breathable air. Air quality checks are required before entering and at regular intervals during occupation. In large excavations with top-down construction mechanical plant can be operating beneath several completed floors and forced ventilation may be necessary to provide fresh air and to expel exhaust fumes from the mechanical plant.

4.2.4 Site accommodation

The Contractor is usually required to arrange connections of services to the site for the duration of the site works. Site accommodation, such as offices, car parking and toilets, are required for the Contractor and for the RSS and for any sub-contractors for the duration of the period of construction. Other areas of the site have to be allocated for medical facilities, for stores for materials and equipment, for testing equipment and storage of samples, for standing plant, for stationary plant such as slurry treatment plant for as long as bentonite slurry is used and areas for temporary stockpiling excavated spoil prior to its export from site. The Contractor mobilises resources, often by letting sub-contracts and manages the site by allocating access and working space for the succession of various activities.

4.2.5 Enabling works

For some work, such as traffic diversions, utility diversions may be necessary. If the deep excavation is located within a street, the diversions of traffic and utilities may be extensive. Some utilities such as water supply and gas mains must be diverted by the owner of the utilities. Advanced works for diversion of major utilities may be ordered before letting the main contract. For some utilities such as cables, The Contractor may be required to excavated and assist the owner of the utility in slewing the utilities out of the way. Sometimes the contractor has to provide new ducts on a new alignment, and the owner diverts his utilities via the new ducts.

In developed areas The Contractor is usually required to conduct a detailed building condition survey for all the buildings that are within an "influence zone" around the site. This survey serves two purposes. The first is to establish records of the actual condition of the buildings to be compared with a later condition of the building in the event of a claim against the Contractor or the Employer for damage to the building. Pre-existing cracks should be recorded, and photographs taken. The second purpose is to check whether the building needs protective work and if protective work has been ordered to verify how the protective works may be carried out. Where protective works have been ordered the Contractor generally is required to put them in place before bulk excavation commences or before a certain stage of progress of the works.

Sometimes a contractor is required to design temporary works. Temporary works can include the ELS system, suspended decking to provide working space for areas of the site, ramps for plant or vehicles to descend to the level of the excavation, staging for offices and other installations above the site where space is restricted at ground level according to the circumstances of the site. Sometimes the whole or part of a site is enclosed or partially enclosed in order to reduce dispersion of sound and or dust to the surrounding areas.

4.2.6 Main work

The main work of a contractor is to excavate to the required lines and levels, to provide the necessary ELS system as the excavation advances and to complete any permanent structures that are required within the excavation. Bulk excavation is a major task. For example, a two-level underground station for an eight-car train requires excavation of about 100,000 to 150,000 tonnes of soil. A main-line terminus station might involve shifting 7 million tonnes of earth and carting the spoil off-site to approved disposal locations. Vehicles exiting the site on to public roads normally require a tyre wash to keep the roads clean and temporary strengthening of the exit point to carry the heavy wheel loads for tens of thousands of crossings or more.

In order to excavate, the contractor provides earth lateral support, which includes any necessary bracing. Any temporary platforms and ramps have to be built and often permanent underground structures are to be built. When the site is located within an open area such as in a road reserve, some of the existing underground utilities may have to stay in place during the excavation. The contractor should excavate carefully to expose the utilities and suspend them with temporary structures such as from the underside of a temporary deck from beams spanning across the excavation from wall to wall or supported on piles. Sometimes a temporary support has to be placed underneath the utilities such as by pipe jacking to lay structural beams beneath the utilities.

Permanent underground structures within the excavation are usually built from simple reinforced concrete. Invariably, the bottom slab is cast on blinding. Top-down construction as described in Section 2.2.2 involves casting parts of the other floors on blinding whereas bottom-up construction requires formwork to cast the upper floors. Columns and staircases generally require formwork. Some underground structures, particularly those located within road reserves, when completed are required to be backfilled to the original ground level, and some utilities might be reinstated and the surface paved or landscaped.

4.2.7 Sub-contractors

In South East Asia sub-contracting is quite common. Some construction companies operate only as management contractors and sub-contract work. Some construction companies still employ direct labour and own or lease their own plant. Specialist skills such as installing instrumentation are often sub-contracted. In Singapore it is required that the monitoring of instrumentation shall be carried out by a company that is independent of the main contractor. Very often specialist grouting, say to improve the properties of the ground, is sub-contracted too.

4.2.8 Site staff

Whether provided by The Contractor or by sub-contractors, a team of construction site staff are required for working on deep excavations. There has to be somebody assigned to the tasks described in his section. The numbers of staff vary according to the size of the project and the roles of the contractor. Typically located on site the contractor needs a project manager, some engineers, programmers, quantity surveyors (QS), land surveyors, works supervisors, safety officers, plant operators, foremen, gangers, labourers and security staff may be direct labour or sub-contracted. The contractor's staff are responsible for the management of the site and programming and management of the construction work, procuring plant, materials and workforce. Safety on site is important and is the responsibility of the Contractor. Public relations are usually conducted jointly with the Contractor and are led by the RSS.

For civil engineering public works, contracts are administered by the Engineer. Procedures for administration are stated in Project Administration Handbooks [14]. The Contractor is responsible for all the construction work. Importantly the Contractor's site staff, whether belonging to the Contractor or to a sub-contractor, provide supervision of the labourers. Supervision implies a continuous activity and should not be confused with inspection. Inspection is normally carried out by the RSS employed on behalf of the Employer. Ultimately the RSS inspect and certify the work that has been completed. However, as construction proceeds the RSS need to inspect partially completed work before it is covered up by subsequent work. Supervision of specialist services may require specialist supervisors. Sub-contracted specialist work, such as ground improvement by grouting is often self-supervised by an engineer who is provided by the sub-contracting grouting company. The setting out for the grouting operations may be conducted and supervised by the Contractor's staff, and erection and pre-loading of steel bracing is often supervised by engineers employed by the Contractor.

RSS are required to supervise the contract and inspect the works in progress and upon completion. Requirements and duties for RSS who are employed on HKSAR Government projects are stated in a manual for deployment of RSS [15]. Requirements of the BA for Qualified Supervision for private buildings works generally are stated in APP-48 [16]. The RSS establishment includes not only the site inspectors but also professionally qualified engineers and QS and sometimes land surveyors. Sometimes the designer has a representative on site to respond to issues related to the design.

4.2.9 Records

RSS in effect are the "eyes and the ears" of employers. RSS review contractor's programs and performance and convene regular site progress meetings with the contractor. The objective of such meetings is to track the completion of the works so that they are completed within the contract period, to anticipate any problems and to address the problems beforehand. The Contractor is usually required to report on progress that has been achieved, and the monitoring results are reviewed. He reports on any incidents and inspections for safety.

Applications for periodic payments, usually monthly, are drafted by the Contractor's QS based on the work that has been achieved and checked by the RSS's QS. When agreed, applications are forwarded to the Employer for payment.

Inspection of the work is an on-going procedure. RSS Engineers and Inspectors are informed regularly, usually daily, on work to be carried out, and The Contractor issues Requests for Inspection so that inspectors have due notice to inspect areas of the work when ready and before they are covered. Reports on inspections are certified by the Contractor's staff and by RSS. Any defects found are reported and have to be rectified. A site diary is maintained by RSS to record the weather, personnel on site, activities carried out, plant utilisation and achievement every day. Such records are crucial to determine the validity of claims and the quantum of valid claims.

By a few weeks before handing over the works to the owner, RSS are required to have inspected the project in its entirety and compiled a list of defects, if any. In the event of major defects outstanding, the hand-over of the site may be delayed for the defects to be rectified. There is usually a maintenance period after hand over during which temporary access may be provided to the Contractor to rectify minor defects. At the end of the maintenance period the works should be free from detectable defects and can be certified as complete and final bills can be paid, bonds and retention monies can be returned to the Contractor. If defects persist final completion may be withheld, and occupation may be affected. Sometimes major defects such as consolidation of weak soils beneath an underground structure could result in subsidence that goes on for years. Such cases can have serious financial consequences as to the extent and cost of remediation and service-ability of the underground facility, performance bonds may be withheld, and substantial claims may arise. Sometimes defects are not recognised until after final completion has been certified and bonds have been released. Contrac-tual provisions usually do not extend beyond a specified maintenance period and claims for damages that appear may be addressed until Common Law. In Hong Kong, in Tort, the time limit is six years after some damage occurs as a result of a wrongful act. In the United Kingdom the time period is also six years, and in other countries the time limit varies from a minimum of two years in Turkey to 30 years in Belgium [17]. Therefore, a prudent employer would be well advised to sustain inspections and monitoring of the works and notify the contractor of any untoward performance such as sub-sidence or cracking as soon as it is detected.

Taking records is important for determining stages of completion, interim payments and final payments. Detailed records of site staff and plant activities and achievements are taken. These are used in compiling valuations and are also important in the event of claims being made for additional or unforeseen work. The Contractor's responsibilities on site are to complete the construction in accordance with the contract, to com-pile as-built records, to clear the site and prepare it to be handed over.

Instrumentation is now very extensive to monitor the performance of deep excavation works both within the site and in the building protection zone around the outside the site. For deep excavations instrumentation and monitoring are very important for the safety of the works and for limiting ground movements so as to prevent damage. As discussed in Sec-tion 3.4, AAA limits are, or should be, set for every instrument, and dif-ferent limits may apply to the same instruments at different stages of the work. In the event that AAA limits are approached a contractor's staff and RSS should review the monitoring records and decide what actions should be taken as appropriate. There are usually so many instruments and frequent readings that data management systems are needed for monitoring data to be readily available to authorised users. User-friendly systems are in use whereby all of the site monitoring data can be accessed via the internet, most of it in real time, and warnings are circulated to responsible persons automatically when instruments reach AAA levels. In Singapore it is a requirement that monitoring of instruments shall be con-ducted by an entity that is independent of the Contractor.

4.2.10 Statutory duties

In Hong Kong and in Singapore, as discussed in Section 3.7, legislation relat-ing to building work requires registration of qualified persons who take personal responsibility for their roles and, according to the severity of any default, registered persons can be subject to a fine, de-registration, or a custodial sentence. In Hong Kong the registered person who coordinates all the work for design and supervision is the Authorised Person (AP). Specialist engineering services supporting an AP are provided by a Registered Structural Engineer (RSE) and/or a Registered Geotechnical Engineer (RGE). Work shall be carried out by a Registered Contractor (RC). In Singapore the requirements are slightly different. There is a Qualified Person for Design (QP(D)) and a Qualified Person for Supervision (QP(S)) and the Qualified Contractor (QC). The statutory duties of registered persons during construction are discussed in Section 3.7. The statutory duties fundamentally address public safety and do not address economy or speed of construction.

4.3 Claims and resolution

4.3.1 Claims

To many people the purpose of having provisions for claims in civil engineering contracts is misunderstood. Whereas for building contracts for structures, the materials such as concrete and steel are manufactured products for which the risk is solely in price fluctuation of basic commodities such as labour, fuel, steel, cement and aggregates. The prices of commodities follow trends in the market. Fundamentally the ground is far more variable than trends in the costs of commodities. When unexpected ground conditions are encountered, there is often a large cost or time implication for work in the ground such as deep excavations and tunnelling. For example, the presence of rock when soil was expected has a considerable impact. Rock requires more effort and can be slower to excavate and generally requires mobilisation and use of different equipment. If a contractor is geared up to excavate in soil and he encounters rock instead he may be obliged to mobilise different equipment. He might also be obliged to mobilise more than one piece of equipment in order to achieve the rate of productivity required to complete on time. An unforeseen ground condition is an important risk that has to be addressed in a contract. Nowadays it is a general practice to identify in a contract what ground conditions are foreseen in order to avoid disputes over the definition of what is foreseen and what is not foreseen, as discussed in Section 3.6.

Many people incorrectly assume that claims are something of a wrong practice. Stories about unwarranted claims probably account for the perceptions that claims are wrong. In civil engineering contracts it is recognised that unforeseen conditions might be encountered, and claims are fundamental to addressing changed conditions. It is a bad practice to make false or exaggerated claims.

Fundamental to a contract is the time for completion. A contract normally has a contract period with a specific date for completion. There are usually provisions for late completion such as the imposition of liquidated damages. These are charges by an employer for late completion, and they are usually specified in a contract in dollars per day, sometimes millions of dollars per day for large projects. If in the event of changed conditions the time to complete the work extends beyond the contracted date for completion, a claim for impossibility to complete within the contracted period would arise. Such an eventuality can be overcome by granting a contractor an extension of time.

The procedure for claiming is as follows. As soon as a contractor identifies grounds for a claim, he should notify the engineer, or the engineer's representative on the RSS. It is important that the Contractor and the Engineer, or RSS, should record and verify any and all relevant site records and actions taken that might be required in order to settle the claim. Initially the Engineer makes a decision as to the validity of the claim and its value. If the Contractor accepts the Engineer's evaluation, the matter is settled. If a contractor disagrees with the evaluation, a dispute arises then the claim requires resolution by a third party.

4.3.2 Resolution

There are several means of obtaining a resolution to a dispute over a claim. These include a court hearing, arbitration, mediation and expert determination. A court hearing follows procedures for Civil Law, and the hearing is in public before a judge who is appointed by the Judiciary. Arbitration follows similar procedures except that the hearings are held in camera and the findings are private to the parties. Mediation is a procedure whereby an employer and a contractor meet to try to resolve their differences, sharing the pain so to speak, at less cost than seeking a decision in law courts or in arbitration, both of which are costly processes. Expert determination is adjudication by an expert on claims that are entirely on technical matters.

Arbitration is a formal hearing with adjudication and an award. When a contract provides for arbitration or when an employer and a contractor agree to arbitration, the two parties agree upon an arbitrator or a tribunal and employ law firms who instruct legal counsel to represent their cases. In the absence of agreement to seek arbitration, the claims may go to trial in the civil law courts in front of a presiding judge. In the law courts there is no choice by the parties over which judge. The judge is appointed by the court. The hearings for arbitration and for trial are broadly similar and follow the Civil Procedure Rules of the United Kingdom [18]. In arbitration and in court hearings engineers play a role. Engineers who were involved in the subject of the dispute may be called as witnesses of fact and present a statement to be submitted under oath and be examined and cross-examined by counsel. Independent engineers may be called by either party to give expert evidence and to tender an expert report. The Court Procedure Rules for Hong Kong, Singapore and the United Kingdom require that the role of an expert is to assist the court and not to advocate the case for one of the parties. When experts are appointed by both parties there may be meetings between the experts in order to identify points of agreement and disagreement. The disagreements then remain for the court to decide. Next the experts are called to give their opinions under oath in response to examination and cross-examination and re-examination respectively by counsel for the two parties. Recently "hot-tubbing" has been introduced whereby the court conducts a concurrent examination with both experts on a given subject. This procedure is aimed at making clear to an arbitrator the essential differences in opinion between the experts and the reasons for the differences so that the arbitrator can then weigh the two opinions. From my experience the procedure seems to save time and adds clarity. In some cases, the court appoints a single expert to give an opinion on matters within the expert's field of expertise.

Mediation is a meeting of the two parties facilitated by a mediator. The two parties, employer and contractor, present to each other the basic elements of their respective cases. They are able to deploy lawyers to present their cases because often there are points of law to be considered. The two parties are also able to call upon experts to address technical points. The mediator does not adjudicate. He is there to keep focus and control the presentations and keep time and may recommend at certain points that the two parties withdraw to see if they can resolve certain points in camera without prejudice. It is entirely up to the two parties whether they want to agree a settlement. Mediation is a quicker process than arbitration taking perhaps one or two days rather than several weeks or months.

Expert Determination is adopted when the two parties identify claims that are entirely on technical matters. An Expert Determinator (ED) conducts an adjudication based on representations from both parties, which might include expert reports. The representations are sometimes written depositions and there then may be no hearing, in which case the expenses of the adjudication are quite modest. A hearing can be conducted if both parties agree, but it adds to the costs. The two parties have to agree beforehand whether the determination by the ED is binding. If the determination is not binding and the case is taken to arbitration or to the civil law courts, there is a risk that the expert determination is used as a sounding out for the technical merits of the respective cases.

References

[1] LC paper No. CB(1)1328/13-14(03) Legislative Council, HKSAR Government. www.legco.gov.hk/yr13-14/english/panels/tp/tp_rdp/papers/tp_rdp0505cb1-1328-3-e.pdf

[2] Moorwood, R., Scott, D., & Pitcher, I. (2008). Alliancing, a Participant's Guide. AECOM, Australia. ISBN97-0-646-50284-7.

[3] Lewis, D. (2016, September 27) Unforeseen Ground Conditions, Paper Given to Society for Construction Law, Hong Kong. www.scl.hk

[4] General Conditions of Contract for Civil Engineering Works. HKSAR Government. (1999) www.devb.gov.hk/filemanager/en/content_188/gf548.pdf

[5] Essex, R.J. (2007) *Geotechnical Baseline Reports for Construction, Suggested Guidelines*, American Society of Civil Engineers, Reston, VA. ISBN 13:978-0-7844-0930-5.

[6] Ip, A.W.C, Lam, E.W.C., & Cheung, P.C.W. (2009) Design and Planning of Lai Chi Kok Transfer Scheme, Hong Kong. *Proceedings of ITA-AITES World Tunnel Congress 2009*, 22–28 May 2009.

[7] New Engineering Contract (NEC): Institution of Civil Engineers, London. www.ice.org.uk/what-is-civil-engineering/what-do-civil-engineers-do/new-engineering-contract

[8] NEC News. (2016) www.neccontract.com/About-NEC/News-Media/Hong-Kong-government-endorses-NEC-as-main-contract.

[9] *Project Administration Handbooks*. (2018, October) Civil Engineering Department, HKSAR Government. https://www.cedd.gov.hk/eng/publications/standards-spec-handbooks-cost/stan-pah/index.html (21 March 2020).

[10] Management Handbooks Rev B-15. (2018) Development Bureau (Works Branch), HKSAR Government. https://www.devb.gov.hk/en/construction_sector_matters/contractors/contractor_management_handbook_revision_b/index.html

[11] WBTC No. 10/97A. (1997) Works Branch Technical Circulars, Development Bureau, HKSAR Government. www.devb.gov.hk/filemanager/technicalcirculars/en/upload/.../wb1097a1.pdf

[12] Construction Site Safety Manual. (2018) Development Bureau (Works Branch), HKSAR Government March 2018 www.devb.gov.hk/en/publications_and_press_releases/publications/construction_site_safety_manual/index.html

[13] *Project Administration Handbooks*. (2018, October) Civil Engineering Department, HKSAR Government.

[14] *Management Handbook for Direct Employment of Resident Site Staff for Public Works Projects*, Development Bureau (Works Branch), HKSAR Government, July 2018. www.search.gov.hk/result?tpl_id=devbii&gp0=devbii_home&gp1=devbii_home&ui_charset=utf-8&web=this&ui_lang=en&query=Management%20Handbook

[15] APP-48 Practice Note for Qualified Supervision of Structural Works and Excavation, Works Buildings Ordinance Section 17, Building Authority, HKSAR Government. August 2019. www.bd.gov.hk/doc/en/resources/codes-and-references/practice-notes-and-circular-letters/pnap/APP/APP048.pdf

[16] Thomson Reuters Limitation Periods by Practical Law, Thomson Reuters, 1st Feb. 2019. https://uk.practicallaw.thomsonreuters.com/1-518-8770?transitionType=Default&contextData=(sc.Default)&firstPage=true&comp=pluk& bhcp=1

[17] Civil Procedure Rules. (2019) Ministry of Justice, United Kingdom. www.justice.gov.uk/courts/procedure-rules/civil

Chapter 5

Current practices, problems and the future

Having described the many activities required for deep excavations, we may ask: How does it all fit together? This chapter examines some current practices, identifies some good practices and some of the instances where matters did not go to plan and what lessons were learned from some problematic projects; it closes with some comments on what lies ahead with respect to deep excavations.

5.1 Current practices

Whereas 50 years ago, whether working in design or in construction, engineers needed to be versatile and resourceful making use of relatively simple techniques, in the last few years the industry has benefitted from many technical advances aimed at assisting engineers to obtain information more reliably, to accomplish more than they used to do and more accurately and to reduce repetitive actions. This section describes some of the current practices that have improved the technology for constructing deep excavations.

5.1.1 Site selection

For public sector and utility projects, site selection may be undertaken at the start of the project. For transportation including railways and roads that can sometimes require deep excavations, there may be options identified for routes to be studied and to be ranked for selection. For specific sites for deep excavations, especially for private sector works, there would be no need for site selection and only the next step of site characterisation is needed. To evaluate potential sites, and for characterisation of selected and specific sites, a study of the landform, previous usage of the site and geological information is generally needed. Aerial photographs have long been a means of evaluation of landform and previous site usage. Modern techniques of remote sensing are more numerous, including satellite imagery, radar and LiDAR [1], and are more versatile. Remote sensing at different frequencies and the use of drones can be obtained or commissioned for project-specific surveys. LiDAR is similar to radar but uses pulsed laser instead of radio waves. LiDAR achieves a better measure of the earth's surface than photography or radar, both of which reflect off the top of dense vegetation whereas LiDAR reflects off the surface of the ground unaffected by vegetation.

Stored geological and geotechnical information is now extensive in many countries and is continuously increasing as data is added to existing databases and more databases are created. The HKSAR Government realised the value of stored geological data in 1980 with the creation of a geotechnical library now with over 164,000 items including periodicals, books, conference proceedings, manuals, standards, codes of practice, geotechnical reports, maps and other references. The library also houses a digital geotechnical information unit where over 400,000 archived borehole logs are stored and are available [2].

Landform, topography, underlying geology and surficial ground conditions are all factors that are evaluated in site selection and site characterisation. Such information can be gathered and provide useful insight about the site or sites before embarking upon GI.

5.1.2 Site investigation and ground investigation

A comprehensive guide to site investigation, including methods of drilling and sampling, equipment used, in-situ tests and laboratory tests can be found in Geoguide 2 [3]. Site investigation obtains information about the site, of which GI is the exploration of the ground beneath the site. Site selection includes an investigation of alternative site locations. Once a site is selected, it is studied usually in more detail but using the same methods that are available for use in site selection. A desk study is carried out on available information form archives, from regional and local geological studies, maps and memoires. When available, archived borehole data from nearby sites or from previous use of the site itself are all useful in understanding the site conditions. Once such stored information has been gathered and assessed, a GI can be planned and carried out. GI can be expensive, and there is usually a budget. It is prudent to plan GI to get the best possible information about the underlying ground conditions within the budget. Whereas for relatively small sites in urban areas a borehole in each corner might be all that an employer will pay for, boreholes can be costly and should not be wasted. Ideally, boreholes should be aimed at specific targets, whether to identify the top of rock across a site, exploring the size of caverns in karst or for other reasons. As the late Sir John Knill [4] once said, "*If you do not know what you are looking for you are unlikely to find it*" (I paraphrase). Time spent in specifying bore holes, to what depth and, sometimes, at what orientation, is time well spent.

Very often GI is based almost exclusively on sinking exploratory bore-holes, recovering samples of the ground for purposes of identification and testing, testing the ground in situ, recording ground water levels and taking samples of groundwater for testing. There are standards for GI and testing [5] in which descriptions of methods are to be found. In addition to drill holes, there are other investigation techniques such as seismic profiling, resistivity profiling, magnetometer and gravity surveys.

5.1.2.1 Boring and drilling

Originally boreholes were sunk by driving robust steel tubes into the ground with blows from a hammer and retrieving samples of soil trapped inside the tubes, or by rotating an auger into the ground and retracting it with soil on the flight of the auger. These methods often resulted in only disturbed samples of soil, and sometimes the recovery of samples was incomplete especially in friable soils because of soil falling out of the sampling devices. The driving method has been improved for weak soils by pushing or vibrating thin steel tubes into the ground and withdrawing them to recover samples of soil inside the tubes. Using a vibrator to drive a thin tube is a good means of recovering continuous samples of very weak soils. The method has limitations. The length of the sampler, typically 5m, limits the depth of sampling, and the method will not penetrate strong ground such as firm clay or granular soils. Thin-wall samplers, up to 1m long, are often used to take samples from the bottom of a larger diameter drilled borehole and this method can be used when very soft soil extends more than 5m into the ground.

Drilling is becoming a more common method for sinking holes. Drilling equipment excavates the ground by rotating and driving a steel casing and removing the soil from inside the casing. In most cases, for deep excavations drillholes are vertical and are sunk to sufficiently far below the intended depth of the excavation in order to identify soils that might influence the performance of the excavation and subsequent underground structures. For investigating specific geological features, inclined drillholes can be adopted. In rock, drillholes can be steered and cored (directional coring) for long distances; they can be steered horizontally or potentially in any direction. At present, drillholes cannot be steered in soil, and directional coring serves no purpose for deep excavations in soil. Inclined drilling and sampling is sometimes used in soil. Usually drilling is vertical and penetrates the ground with a rotating steel casing with a cutting edge at the base. The steel casing cuts the soil and serves to prevent the sides of the drilled hole from collapsing. Water is commonly used to flush drill-holes during drilling, to aid the drilling and to remove the loosened soil or fragments of rock. Flushing with foam [3] instead of water assists in achieving more complete recovery of samples of soil, especially friable soil and soil that is mixed with fragments of rock material, such as cobbles and boulders that require coring. Flushing with water results in poor sampling of mixed ground, comprising soil and rock. While the rock is cut slowly the soil usually gets washed away by the flushing water and only fragments of rock are recovered as samples. Moreover, rock fragments can get displaced during drilling and move around whilst being cut and only irregular fragments of rock material are recovered.

5.1.2.2 Sampling and logging

The simplest form of identification of the ground is by the driller examining the returned flushing water and the recovered parings of soil and rock fragments. These particles of spoil are highly disturbed, and a description of them is not very reliable because the in-situ condition is completely broken up and fine-grained particles that are suspended in the flushing medium may not be taken into account. Samples are retrieved, sometimes continuously and sometimes intermittently. Sampling can be undertaken by driving a tube into the ground as is adopted when boring. These samples are more complete than the fragments that are suspended in drilling fluid; and the grading, the relative quantities of particles of different sizes, such as clay, silt, sand and gravel, can be gauged visually. A commonly used sample tube is adopted for Standard Penetration Tests (SPTs). SPTs are often carried out to recover samples at regular intervals of depth as a drillhole is advanced. The SPT involves driving a small standard tube sampler into the ground below the bottom of a drillhole using a standard hammer weight and drop. The number of hammer blows is counted for three successive penetrations of 150mm each. Blows to penetrate the first 150mm are recorded but not included in the total blow count because the soil might have been disturbed by the drilling process. The numbers of blows for the second and third stages of penetration are added together, and the result is the SPT "n" number. The SPT number gives a measure of the resistance of the ground to penetration and can be correlated to strength of the soil and its stiffness. In addition, the hollow tube contains a sample of the ground that can be used to identify the type of ground at the level of the test. Samples are usually considered to have been disturbed because the sampler has a thick wall, but often in firm ground the retrieved sample is only disturbed a little. Such samples are usually smeared on the outside, and the fabric of the soil can be examined when the sample is split open. SPTs are intended for use in clay, silt, sand and gravel soil. The sharp cutting edge on an SPT sampler is damaged if it hits hard material such as rock or very dense gravel.

When undisturbed samples are required for testing, thin wall steel samplers are used for weak soil. In firm soil, taking a sample requires cutting the ground. A popular tool for this is a Mazier retractable core barrel, which is designed to recover undisturbed sample from very soft soil and can be used for firm soils. It comprises three concentric tubes. The outer tube is a strong casing with external diameter 101mm and with strong teeth around the bottom edge for cutting firm soil and rock particles such as gravel and cobbles. Inside there is a thin steel tube "core lifter" with a sharp cutting edge that is used to trim weak soils. Inside there is a PVC core liner, which is removable and serves to retain the sample of soil for transportation and storage. The recovered sample is 74mm dimeter, which is a standard size for laboratory triaxial testing apparatus.

Friable soils are very susceptible to disturbance when they are sampled and require special sampling if the in-situ properties are to be preserved and a sample is to be retrieved intact. Ground freezing is one method, although water expands when it freezes by 10% so the sand is disturbed by this means. Injection of chemicals into the soil to bind the sample together is another method. In-situ testing by SPT or CPT is probably the most common test for sand.

Descriptive logs of the drillholes are prepared on site by the driller based on inspecting particles of soil suspended in the returned flushing liquid and from the ends of the SPT samplers. The driller can only report what he sees: mostly disturbed spoil and the ends of samples that are confined in tubes. The user of driller's logs should realise that the driller has not seen undisturbed ground, but his observations can be useful because they are based on ground that has been disturbed; such observations can help one to understand how the properties of the ground can be changed by mechanical disturbance and by additional water. Sometimes the driller is assisted or supplanted by a field geologist who takes responsibility for field logging.

It is customary for the design staff to review the driller's logs or the field logs, to open and inspect samples and to make use of the results from laboratory tests to prepare fresh logs. Field logs are, of necessity, mostly based on the examination of disturbed soil. By opening samples, an engineer or geologist can inspect more of the soil and in its undisturbed state; laboratory tests can aid in differentiation better definition of the characteristics of the soil. For example, distinguishing between silty clay and clayey silt from a suspension in flushing water is not easy, whereas laboratory tests can measure the exact percentage of clay and silt in a given specimen.

5.1.2.3 Profiling

Drillholes are mostly sunk from the ground surface. Many are vertical and some are inclined. Therefore, they only sample the ground at one location. Several drillholes are usual for a site, and interpolation is used to estimate the ground conditions between drillholes. There are several methods of geophysical profiling the subsurface layers and variations of types of ground between drillholes including seismic refraction, ground penetrating radar, micro-gravity and resistivity surveys.

Seismic refraction profiling has been in use for at least 50 years. The method depends on different velocities of small seismic waves in ground with different densities. The denser the ground is, the faster the seismic velocity. The technique was first developed with simple equipment used on land. On-land seismic refraction surveys were conducted along lines to obtain geological profiles by striking a plate placed on the ground to send a small seismic wave through the ground to receivers, called geophones. The first arrival time of signals at successive geophone locations depends on a seismic wave finding the fastest path. Initially the fastest path is directly along the surface of the ground at a relatively slow speed (surface wave). At a certain distance the seismic wave travels faster by going down to a denser layer where the seismic velocity is faster and then reflecting back up to a geophone (reflected wave). At a greater distance, the travel time is faster by going down a lower layer, then travelling quickly along that layer and back up to the surface as illustrated in Figure 5.1.

In the past, data was computed and interpreted manually; this required skill and experience. Resolution is feasible between ground of sufficiently different densities. Differentiation between soil and rock is relatively clear depending on the depth to the interface, the singularity of the interface and correlation with drillholes along the profile under investigation. Short surveys can be conducted by generating a signal wave by a hammer blow on a steel striker plate. The distance and resolution can be improved by conducting repeated blows and adding the received signals for the same array of geophones. Explosive detonators can be used to generate more powerful signals, which can be detected at greater distance and achieve a correspondingly deeper penetration if needed. Using a larger seismic source, arrays of geophones can be set in different directions from the

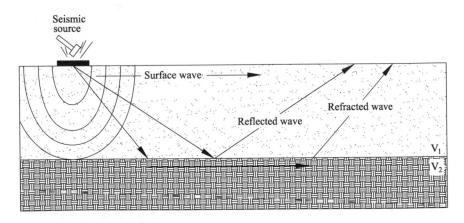

Figure 5.1 Geophysical survey.

source, and more information is obtained over a wider area for each test than can be achieved with a low-energy hammer blow.

Seismic surveys over water are faster than surveys on land. A boat is used to tow both a source and receivers. The source sends out seismic waves repeatedly and the receivers feed successive records of return signals to a computer on board the boat. Marine seismic surveys can survey large areas in only a day, and the profiles can be conducted on a grid of orthogonal paths thereby covering whole sites in both directions. The interpretation of continuous surveys is usually based on detecting reflected seismic waves. Interfaces between different types of ground reflect seismic waves when the underlying material is denser than the overlying material. Interpretation of reflected waves depends on having several boreholes on, or very close to, survey lines for correlation. In the early days, interpretation by hand was time-consuming and subjective requiring skilled interpreters. Geophysical surveys now are more automated and less dependent upon skilled interpretation, and other ground-profiling methods have been developed [6], some with more specialised applications and with computer-based means of data gathering, storage and interpretation.

Ground penetrating radar is useful for locating subsurface structures and can be used to profile rock head. The method relies on reflection of a radio frequency wave from a transmitter back to a receiver, which is usually with the transmitter in one compact unit. The equipment is quite mobile with small hand-held units or units that can be towed behind a car, but it has a limited depth of penetration.

Micro-gravity surveys track minute changes in the gravitational field, as small as 1 part in 1 million, and are used mainly to detect cavities in the ground or changes in density of the ground. The largest responses come from cavities in rock. The device is a single compact and easily portable receiver unit [7].

Electrical resistivity surveys measure electrical resistivity of the ground across an array of electrodes spread along a traverse for profiling. The electrical resistivity varies with type of soil and water content and the method helps to profile transitional boundaries between sub-surface layers [8].

5.1.2.4 In-situ testing

Recovery and transportation of samples can disturb their properties. Testing in situ does not rely on sampling, but it can be affected by disturbance of the drilling process that formed the drillhole. As described in Section 5.1.2.2, a very common in-situ test is the SPT. Whereas the SPT method generally results in a disturbed sample, the n number relates to penetrating more than 150mm below the bottom of a drillhole in an attempt to test undisturbed ground.

A similar approach is adopted with a field vane test whereby a small vane is pushed into ground below the bottom of a drillhole in an attempt to reach undisturbed ground. The vane is rotated and shears the ground on a cylindrical surface. The torque required is recorded to measure the shear strength of the ground. The method is suitable in soft clay and with a smaller vane in firm to hard clay.

Pressuremeters press the ground sideways and measure the displacement, which can be used to compute stiffness and strength. Fifty years ago, a cylindrical pressuremeter was designed by Louis Menard [9], and Menard Pressuremeters are still very much in use. The method is to lower a cylindrical rubber bag down a drillhole and inflate it so that it expands the hole and compresses the ground around the drillhole. The displacement is measured by the volume of injected fluid, and the pressure is recorded. These parameters can be used to estimate the stiffness of the ground around the perimeter of the drillhole at that depth and its strength. Expanding a cylindrical cavity in this way results in tensile circumferential strain around the perimeter of the cavity and the ground immediately in contact with a pressuremeter reaches its shear strength at a small displacement. The stiffness that is computed by initial loading and displacement can be a combination of yielding for a short distance and elastic strains further away. Therefore, the computed stiffness is less than for undisturbed ground. By unloading and re-loading to the same pressure, the displacements are almost reversible and computed stiffness is usually greater than that obtained from the initial loading. The unloading and reloading stiffness is often adopted as the elastic stiffness of the ground.

Menard's pressuremeter tests the exposed sides of a drillhole that could have been disturbed by the drilling. Two other tools aimed at testing undisturbed ground have been developed. One is a self-boring pressuremeter, which has a cutting tool at the base and a flushing mechanism. It operates like a small drill and is used to drill into the ground below the bottom of the drillhole while advancing the pressuremeter tool with minimal clearance to minimise the disturbance to the ground that surrounds the tool. The pressuremeter also expands radially and the data can be interpreted in the same way as for Menard's pressuremeter. Test procedures for cylindrical pressuremeters are given in BS 5930 [10].

Another tool intended to test the ground in situ with little disturbance is the flat plate dilatometer. This instrumented pair of thin stainless steel blades comprises two flat plates with an expandable steel membrane around the edges. The blade is driven into the ground below the bottom of a drillhole and inflated, thereby driving the two plates apart. The pressure and the displacements are recorded, and the results can be used to determine the in-situ horizontal stress, the strength and stiffness of the ground that is immediately in contact with the plates. The test procedure is described in ASTM DD635-15 [11]. Whereas expanding cylindrical cavities result in very quick development of large hoop strains around the expanding cavity, and in plastic deformation developing very quickly, by using flat plates, there is sufficient elastic deformation during a plate test to determine the elastic stiffness of the ground more accurately before the onset of plastic deformation.

Cone Penetration Tests can be carried out from the ground surface without a drillhole and provide an almost continuous result by taking measurements of the resistance at the tip and shear on the sides of an instrumented probe that is pushed into the ground. The readings are taken at intervals of 3cm and can be processed automatically. Ground water pressure is also measured, sometimes on the tip and sometimes on the shaft of the probe. Depending on the type of soil, pore water pressures are developed as the probe is pressed into the ground, and the three sets of data, tip resistance, shaft resistance and pore water pressure changes can be interpreted to determine strength, stiffness and type of soil with a fair degree of reliability; descriptions of the test procedure are given by ASTM DD635-16 [11]. An interpretation of the data is described by Robertson (2009) [12].

5.1.3 Laboratory testing

Current practices were well established many years ago. There have been few developments other than in storing and transmitting data. Tests and test procedures are set out in standards [13]. There are classification tests and tests for mechanical properties. Classification tests serve to identify and classify the type of soil, such as clay, silt, sand or mixtures thereof. Classification tests include the following:

- Particle size determination, by sieving for sand and gravel and a hydraulic method based on rates of sedimentation for clay and silt.
- Moisture content.
- Atterberg Limits for plastic soil such as clay, Liquid Limit (the moisture content at which the addition of further water turns the soil into slurry) and Plastic Limit (at which point any less moisture renders the soil non-plastic).

By correlation, the Atterberg Limits combined with the natural moisture content give an indication of the shear strength of fine normally consolidated grained soil.

Material property tests determine strength and compressibility and include:

- Laboratory vane test to measure shear strength of weak soil.
- Fall cone test to measure the strength of weak soil.
- Shear box test to measure the shearing rupture strength confined to a horizontal plane when a specimen of soil is also subjected to vertical confining loads. This test is not common now.

- Triaxial compression test, conducted on a cylindrical specimen placed vertically inside a cell with a confining fluid pressure and subjected to increasing axial loads. This test can be drained or undrained and is commonly adopted to measure shear strength and compressibility, both volumetric compressibility and shear stiffness.
- Compressibility, in one dimension, requires drainage and is commonly called consolidation. The test apparatus is called an oedometer and subjects a thin circular specimen placed flat and loaded or unloaded with drainage permitted.

Triaxial and oedometer tests operate at and report strains of the order of a quarter of 1% upwards. These are relevant to situations in the field such as when a retaining wall moves by one hundredth of its height. However, a 30m deep wall moving by 1% of its depth moves 300mm, which in most urban locations is too much to allow. An allowable movement might be 30mm, which is one tenth of 1% of the height of the wall. It follows that precise estimates of movement for sensitive locations should make use of the stiffness of soil at small strains, which is generally of the order of four to ten times greater than, say, at 1% strain. Dynamic loading in a triaxial cell using Bender Element gauges and the Resonant Column Method measure stiffness of soil at very low stress levels. Bender elements generate and measure shear wave velocity in a specimen of soil. In the Resonant Column Method, the frequency of an input oscillating force is adjusted to obtain the frequency at which resonance occurs. From this, shear elastic modulus and damping properties can be calculated. When designing protection of sensitive buildings, the limiting strains are typically very small and the stiffness of soil at such small strains is generally several times higher than stiffness measured by conventional triaxial tests.

5.1.4 Computers

The biggest changes in design methods have been brought about by the use of computers. What use to be computed by graphical means on a drawing board, or on hand-written pages using logarithm tables and a slide rule as a check, can now be performed much more quickly on a computer. Moreover, far more elaborate calculations can be carried out in very short times. The use of computers for designs is discussed in Section 3.1. They are also used for drawing. Drawings produced by computers are mechanically perfect and easily reproduced, and digital files can be transmitted. In the 1970s, my first experiences included having details drafted by hand in Manchester and Sydney, and then tracing-paper copies were flown to Hong Kong overnight to be checked and corrected locally. Many design processes are repeated, and computer-based spreadsheets have long been used for basic calculations and for compiling standardised calculation sheets from the output of more elaborate computer programs.

Preparing programmes for a succession of inter-related activities by hand was tedious, and up-dating programmes meant almost starting all over again. Programming using computer software is not only quicker but includes inter-dependencies and updates and optimises in response to relatively simple commands.

Computers store and give access to massive libraries of information including standard details for drawings, survey data, records, correspondence and anything that is put in it. The dream of a portable office is a reality. I can carry a portable computer anywhere, plug it into an office intranet or sign into Wi-Fi and draw on clouded information. For each project, there are many design files for correspondence, analyses, calculations, drawings and photographs, and there are many more files generated during construction, containing programmes, method statements, site memos and photographic records. One very useful application of databases is monitoring records during construction. For large deep excavations, hundreds of instruments may have been read frequently throughout the contract period. Information technology makes such databases readily accessible to authorised users via the internet on an inter-active basis at any time, as discussed in Section 3.4.

5.1.5 Risk

One of the biggest changes of thinking within the industry during the last few decades has been the means of addressing and quantifying risk. Geological risk and its mitigation are addressed in Section 3.6. Fifty years ago, codes of practice recommended factors of safety whereby live loads, such as traffic loads, were prescribed, but self-weight for retained soil and values for strength and stiffness of soil were determined by the designer. Often average strength was used for soil. Many engineers recognised that mechanical properties of soil are very variable and geological profiles in the ground are variable. Compared to the variability of soil type and properties, the degree of sampling was statistically not at all significant. In those days, a designer was required to exercise considerable judgement.

One of my early tasks as a trainee was to compute wave forces on a platform in the North Sea. I did this for a one in 100 years wave and applied some factors of safety. I then checked the effects of a one in 1,000 years wave and evaluated a total collapse. The less frequent wave was overwhelming. A comparison with stability calculations for a retaining wall using average strengths and calculations using minimum measured strength was not so disastrous but was worrying, and I learned that colleagues were adopting moderately conservative parameters when they considered that they were warranted and not average parameters.

Risk is now appreciated in many walks of life. For human well-being, health and safety regulations now apply. One is not allowed to enter a worksite without being properly briefed and kitted out with Personal

Protective Equipment. While inspecting the basement for Edinburg Tower, Hong Kong, in 1978 I trod on a piece of wood and two nails penetrated my rubber boots. No protective steel-plated boots were to be had. A carpenter cauterised the wounds using three matches. Workers drilling rock at the bottom of hand-dug caissons could not see for all the respirable fine granite dust and refused to wear full-face respirator masks because they were too hot. Within a few years, the death toll from pneumoconiosis, caused by inhalation of fine dust, amongst caisson workers in Hong Kong rose to above 3,000 as I recall. Few of them reached the age of 35 years. As a consequence, the use of hand excavation for caissons has been strictly controlled to the extent that HKSAR Government imposed a ban on hand-excavated caissons [14].

For a designer, the principal technical risk is dealing with uncertainty as to the ground conditions. The methods of mitigating this risk are discussed in Section 3.6. The procedure is to investigate the site, gathering as much information about the ground conditions as possible. Next is to conduct a GI by exploratory drilling, sampling, testing and profiling between drillholes. Of necessity, the GI does not investigate all the ground. For example, samples are often taken at 74mm diameter, and drillholes at the design stage might be 10m apart. In such a case, each drillhole recovers less than 0.005% of the ground, less than one part in 10,000. In assessing, the types of soil present beneath a site and their distribution require judgement and experience. Drilling can disturb the ground, and sampling and transportation can disturb samples. In most cases, disturbance weakens soil, and adopting the properties of disturbed soil can be conservative. However, some loose soils when disturbed can become denser and correspondingly stronger and stiffer than undisturbed soil. Modern codes of practice such as Eurocode 7 [15] recommend not only taking a moderately conservative evaluation of the properties of the ground but also to apply partial factors on the properties of materials in order to allow for the variable characteristic of the ground.

Loadings to be considered in design and applied to engineered structures, such as underground structures, and loadings on roads adjacent to deep excavations are given in regulations and in codes of practice. Modern codes of practice also apply partial factors on the loadings in order to allow for accidental overload.

5.1.5.1 Factors of safety

Deterministic methods of design as used previously used unfactored loading and unfactored strength and required a single factor of safety (FoS) as a ratio of resisting effects divided by disturbing effects. Eurocode recommends two partial factors, one for material properties, PFm, and one for loading, PFl. For simple structures there is little difference and no difference at all when the partial factor that are chosen such that their product is equal to the original FoS, (when PFLI/PFm = Fos). For deep excavations with ELS, there is a significant change. As discussed in Section 3.10, the net loading on an embedded retaining wall is calculated as the difference between the loads from the retained soil minus the resistance from the soil in front of the embedded part of the wall. If the strength of the soil is factored down, the loading from the retained soil is increased and the resistance from the front of the embedded part is reduced so the neutral point where the two pressures are equal is lower when the strength is reduced by a partial factor than when the strength is not factored. Using factored material properties results in a much deeper zone of applied loading than using unfactored strengths. In effect, using partial material factors provides for strengthening retaining walls deeper than would be determined from using unfactored strengths and is consistent with the risk of the ground being weaker than was assumed for the design. In effect, using partial factors is a method of mitigating risk of the ground being weaker than assumed and the ELS being overloaded.

5.1.6 Control

Authorities in Hong Kong and Singapore exercise control over building works primarily to reduce public safety risks. Decades ago, legislation was introduced to establish standards of good practice by registration and discipline of qualified engineers when necessary. In Hong Kong, the Buildings Ordinance was initiated in 1955; it has been updated many times [16]. In 1955, very few projects could be classified as deep excavations. In 1972 and 1976 there were two multiple fatality landslides that led to the HKSAR Government establishing geotechnical control initially focused on slopes but later broadened in scope to include deep excavations. Scrutiny of plans by an independent authority before granting consent for construction is aimed at reducing human error. The collapse of a deep excavation at the Nicoll Highway in Singapore in 2004, as described in Section 5.2.3.2. suffered from several human errors. In response to the collapse, the control of building works, exercised by within the LTA for underground railways in Singapore, was resumed by the BCA.

5.1.7 Monitoring

In the past, monitoring of instrumentation was labour intensive. Persuading a young lad to hold a levelling staff vertically and still on a rainy cold day in Birmingham whilst wiping one's own eyes to adjust and get a reading with a level or theodolite required persistence. Data had to be written up by hand and copied. At least Hong Kong proved to be warmer. However, the density of development around urban renewal sites requiring many monitoring points meant that monthly monitoring reports of daily readings for a deep basement for an 80-storey office building in Hong Kong were transferred in four number 100mm wide-lever arch files. After a year, the files if placed side by side would have occupied 146m of shelf space, and mining into that hard copy database was tedious. Now many instruments are electronic, automatically read at hourly intervals; they can be accessed anywhere that is accessible to the internet.

Automation only does what it is asked to do. An alert surveyor can notice something going adrift. An automatic settlement monitor can report a subsidence of, say, 30mm. A human might notice that a grass verge is very soggy and might suspect a leaking water main before it becomes a flood. Information technology is advancing rapidly. Detecting a leak in an instrumented water main no longer requires a report from an observer. In the past, a displaced instrument might have been noticed only by an astute member of staff; now artificially intelligent instruments, such as Utterberry [17], talk to each other, giving warnings if one is displaced and then resetting its location when replaced. Automation takes out the human error in manual readings and transcription. Real-time monitoring provides up-to-date information so that one can respond quickly to sudden events, such as a ruptured water main; however, in many cases changes in monitored data for deep excavations progress gradually.

Automation makes data readily available. One does not have to plough through lever-arch files. One can ask for plots of results versus time or plots of one reading, such as lowering of the water table, versus another such as the settlement of the ground near the piezometer or standpipe, which would take ages to do manually.

In 1935, Karl Terzaghi, then a professor at University of Illinois [18], consulted on the Chicago Subway deep excavations; his assistant, Ralph Peck [19], was on site acting as his eyes and ears. Besides being a brilliant applied mathematician, Terzaghi was a practical man. He realised the uncertainties about ground conditions and its performance, and he asked Peck to report on how excavations behaved so that if performance was not in accordance with expectations, he could do something about it. From those humble beginnings, the Observational Method came into being, as discussed in Section 3.5. The procedure is to make a prediction of what will happen. Modern computer methods facilitate predictions of displacements in detail within and around the excavation for successive stages of excavation and construction of the underground structures. There can be as many as 50 stages. Predictions can be made for settlement of the ground at any point, for the amounts of lowering of ground water pressures, for the horizontal movement of the ground at any point and in particular at vertical profiles of perimeter retaining walls. Specific predictions can be computed for any instrument for each stage of the excavation. As discussed in Section 3.4, AAA levels should be determined for every instrument, and monitoring during construction can be compared with the respective AAA levels. If the expected performance is not achieved, then the methods or the design can be adjusted accordingly if necessary or desired.

An important part of the Observational Method is to have contingency plans drawn up in case the performance is not as expected, and changes need to be made. Automated monitoring greatly facilitates this method. The AAA values can be stored in the system for every instrument for all respective stages of the excavation. Warnings can be delivered automatically, and real-time monitoring can provide needed information and save time when making decisions.

5.1.7 The devil is in the detailing

It is commonly held in the industry that failures seldom occur in structural members but often occur in the detailing of the connections. Although there were many contributory factors, mechanical failure commenced at strut-to-waling connections initiating the collapse of the Nicoll Highway in Singapore as described in Section 5.2.3.2.

In the interests of efficiency, designers use computers extensively. Complicated programs can compute many stages of deep excavation with interaction between various structural elements and the ground. Time is money, so to speak. The cost of an engineer's time to design a connection, say a bolted connection, well exceeds the cost of the bolts. Libraries of details are available in many design offices to call up and depict for construction. Diligence requires that the appropriate detail be adopted, just as diligence is required to ensure that the detail is properly constructed or assembled on site. The trend within the industry is large projects, with computer–based assistance, with less time available and less manpower assigned to the project. These factors mitigate against taking time to check details and making sure that the detail is the correct one for the job.

Attention to detail is important. A good detail is a fail-safe one. For example, a web of a beam that is bolted to a single plate connection relies on bolts subjected to shear. If the bolts fail and the beam drops, the beam could fall, but if the beam is restrained laterally, it will drop only until it rests on the plate and will not totally collapse. A double plate connection on either side of a web would support a beam even with no other restraint. It is important to pay attention to details making sure that if a connection fails the system does not suffer a total collapse

Waterproofing is usually required for underground structures, and waterproofing is not always waterproof. Sequential construction leads to joints or overlaps in the waterproofing and taking waterproofing materials around corners can be tricky. Waterproofing does not necessarily have to be 100% during excavation. Some seepage can often be tolerated. When ground support is provided by diaphragm walls, the reinforced concrete panels are generally of sufficiently low permeability. Joints between panels often leak. Joints between panels are often detailed to reduce seepage by incorporating a joggle and a PVC water-stop. However, water-stops must be continuous; otherwise, water paths remain. One such water path is illustrated in Figure 5.2.

Details can be over-engineered. About 30 years ago in Taipei, there was a standard joint detail for diaphragm walls. It was intended to make lateral steel reinforcement continuous passing through the vertical joints between wall panels. In order to cast a first panel, a stop-end was devised to allow steel reinforcement starter bars to stick out laterally through the stop-end and a PVC sheet was incorporated to wrap around the final joint. During excavation for a second panel, these starter bars were an obstruction and were frequently displaced or damaged and the PVC sheet was wrapped around them. Bent bars and a wrapped sheet obstructed the overlapping of the steel bars from the next panel, and the joint was a mess of tangled reinforcement, PVC sheeting, mud and poor concrete. The detail proved to be worse than a plain stop-end and sealing leaking joints was a nightmare.

In the first office where I worked there was a draughtsman of whom it was said *"If Harry can draw it, it can be built, and it will work."* Such skills, honed by years of practice and the ability to think in 3-D when working on drawings, are seldom encountered nowadays; modern skills

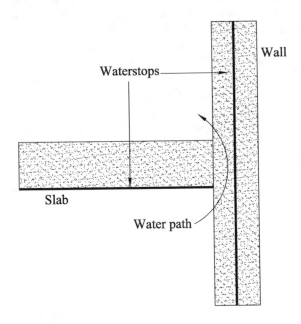

Figure 5.2 Water proofing.

tend to comprise dexterity on a computer's keyboard. Whether or not imperfections such as a pathway will remain in the waterproofing system and be noticed depends on the operator of a computer who might have very little on-site experience.

Waterproofing of underground structures is important not only to provide dry conditions for use but also because any degree of seepage of ground water that evaporates in the underground space runs the risk of leaving behind evaporites, residues from salts dissolved in ground water. Even minute quantities of salts can build up on the inside surface of concrete as the water evaporates, and the residues can become aggressive over time.

Pozzolana concrete [20] (making use of volcanic ash and lime to make cement) was invented by the ancient Greeks and extensively used by the ancient Romans some 2,000 years ago. Several examples of pozzolana concrete survive to this day. Concrete was reinvented in the mid-nineteenth century in an attempt to simulate Portland limestone by firing a mixture of lime and clay. The patent for Portland Cement was obtained by Joseph Aspdin in 1824 [21]. The cement mixed with crushed rock sets as concrete and with various additives has a range of properties, such as high strength, waterproofing and workability. Ordinary Portland Cement (OPC) is now used for concrete, for mortars and for grout.

The effects of pozzolana were also rediscovered, and fly ash, also known as Pulverised Fuel Ash, from thermal power stations provides a ready source of pozzolanic material, which, when added to OPC, improves its performance under wet conditions and improves resistance to chemical attack. There are many additives being marketed to improve the properties of concrete, such as reducing permeability to water and improving resistance to chemical attacks as well as accelerators to harden the concrete more quickly, retarders to delay the hardening and plasticisers to improve the flow of concrete when pumped.

All concretes are characteristically strong in compression and relatively weak in tension. In order to improve the tensile capacity of structural members, concrete has been reinforced by steel rods embedded in the concrete. Steel is strong in tension. Concrete is brittle and crushes with no residual strength, whereas steel is ductile and generally increases in strength after the first onset of yield. Steel-reinforced concrete becomes ductile when the content of steel reinforcement is sufficiently low that the steel yields before the concrete crushes. Characteristically, brittle failure can lead to collapse whereas ductile failure has a reserve of strength and designing with ductility is an added margin of safety such that after the initial failure of a reinforced concrete structural member the member remains ductile with a reserve of strength mitigating against total collapse.

Unfortunately, steel can corrode especially when subjected to alternate wetting and drying, to salt water and to some other contaminants. Corrosion of steel is expansive, and the surrounding concrete can crack and spall thereby exposing steel reinforcement and permitting the rate of corrosion to increase. Corrosion wastes the steel. The steel reduces in cross-sectional area, and the reinforcement becomes weaker and weakens the structural member of reinforced concrete. Some chemicals present in the ground, naturally or by contamination, can also attack concrete. Particularly adverse conditions occur in sabkha areas located near the desert where the soil is rich in evaporites. Protection of steel reinforcement from corrosion is important. Sulphuric acid or sulphate ions that are sometimes found in contaminated ground rapidly attack OPC. It is important when investigating a site for underground structures that the chemistry of the groundwater be checked for any chemicals that would attack concrete or steel reinforcement. Steel reinforcement can be protected by galvanisation, or it can be replaced by stainless steel bars or glass fibre bars. Glass fibre reinforcement is often used for soft eyes in diaphragm walls that will be cut out later. Soft eyes are very useful where Tunnel Boring Machines (TBMs) are planned to break through diaphragm walls into deep excavations when the excavation of approaching tunnels at the end walls of underground railway stations is complete. Tunnel-boring machines can cut glass fibre reinforcement rods more readily than they can cut through steel reinforcement.

For some structures, steel fibres can be used as reinforcement in concrete. Steel fibres can be readily used with sprayed concrete; this combination is often used to quickly apply support to an excavated slope surface of soil to retain it. Steel fibres are used increasingly for segmental tunnel lining but so far have little use for permanent underground structures.

5.1.8 Testing

As is the case for any manufacturing process, testing during construction is very important. Sources of materials should be tested to ensure compliance with the specification. Completed work may be tested for compliance. Testing is also important to verify assumptions and in the design or planning of the project.

For deep excavations below the ground water table, testing the effectiveness of groundwater control before excavation below the water table is very important because any inadequacy of ground water control could adversely affect the works and prove very difficult to remedy. For deep excavations in soil below the water table, ground water control is generally needed to keep the water out from the excavation and prevent subsidence of the surrounding area. Most cases of deep excavations below the ground water table require a control of inflow of ground water through the ELS system and an impermeable cut-off below the excavation to prevent inflow of ground water beneath the retaining walls.

In Hong Kong in the 1970s at the start of construction of the Mass Transit Railway (MTR), as the result of inflow of ground water into excavations for underground railway stations in the Mong Kok District, there was subsidence of the surrounding areas and distress to nearby buildings. The BA imposed limitations on degree of drawdown of ground water outside deep excavations [22]. This requirement has led to the adoption of full-scale pumping tests for deep excavations prior to excavating. In most cases of deep excavations in Hong Kong, there is a peripheral diaphragm wall intended to be deep enough to provide a hydraulic cut-off or a grouted curtain beneath the bottom of the diaphragm walls. Wells are sunk inside the site below the elevation of the final excavation. Water is pumped from the wells and the water table within the site is lowered to beneath the lowest elevation of the excavation. The ground water conditions outside the site are monitored for compliance with a limitation on lowering the ground water outside the site and settlement of the ground. A permit to excavate is not given unless the results of the pumping test meet the requirements.

Conducting a full-scale pumping test is prudent. If excavation proceeds and the cut-off is found to be insufficient, remedial work can be difficult. Once water starts to flow into the excavation stemming the moving water by grouting when ground water is flowing is difficult. Grouting is best carried out in static ground water; otherwise, the grout is diluted, and it can become ineffective or displaced or even wash away. If joints between diaphragm wall panels are open water and sometimes soil can pass through, causing excessive settlement outside the diaphragm wall.

Integrity of diaphragm wall panels is required to exclude water from the excavation and for structural capacity. A number of problems can occur when forming diaphragm wall panels as follows:

- Out-of-tolerance vertically
- Gaps between panels
- Incomplete concreting, honeycomb
- Insufficient concrete cover to reinforcement
- Debris entrained within the concrete or at the base
- Segregation of the concrete
- Over-break during excavation leading to excessive thickness of concrete

Sources of such problems include mechanical operation of the excavator and obstructions including jammed stop-ends from an adjacent panel, insufficient cleaning of debris from the bottom of the trench, lack of control of the properties of the slurry and properties of the concrete and improper placing of the concrete.

Integrity testing can be conducted on completed wall panels. Sonic coring is a procedure whereby vertical PVC pipes are cast inside the panel and a sonic transmitter is lowered and raised in each pipe in succession with receiving probes in other pipes. The transmission of the sonic signal is affected by the density of the concrete between the pipes; any voids or low density can be identified by a procedure called tomography (a procedure creating two- or three-dimensional imagery). Tomography can also be used to scan between boreholes to assist in profiling the subsurface conditions.

For checking the base of a wall panel, physical coring can be carried out vertically through a panel, either from top to below the bottom of the concrete or from the bottom of a cast-in duct. The depth to the bottom of the concrete and the material beneath the bottom of the concrete can be checked. Coring from the top of a wall panel to the bottom also obtains samples of concrete core to check the density and strength of the concrete.

5.1.9 Resolution of disputes

As discussed in Section 4.3.1, claims within a contract should be addressed at first by a decision from the Engineer. However, if the decision is not accepted and a dispute arises requiring an adjudication, recourse is arbitration or a court hearing under civil law. Arbitration can be prescribed by a contract including the location of the arbitration since the procedure varies depending on the country and on the law of the country. If arbitration is not specified in a contract, the two contracting parties may agree on arbitration; otherwise, the dispute can be taken to a civil court for resolution by a court-appointed judge. In arbitration, the parties can choose but must agree upon the selection of the arbitrator, or tribunal. Arbitration is held in camera and is private to the two disputing parties. The decisions are binding, subject only to limited appeal on legal procedures. For this reason, many arbitrators are lawyers. In American arbitration law there exist a small but significant body of case law that deals with the power of the courts to intervene where the decision of an arbitrator is in fundamental disaccord with the applicable principles of law or the contract. However, this body of case law has been called into question by recent decisions of the Supreme Court. The civil court hearings are public, and the decision of the presiding judge can be taken to appeal.

Arbitration and civil court hearings can be very expensive and time consuming. Various methods have been advanced in to accelerate and reduce the costs of dispute resolution. One method is to set strict time limits for representations from both parties. Another method has been dubbed "hot tubbing" whereby experts from both parties are called to give evidence in response to questions put by the court. Having experienced hot tubbing I have found that it was effective in shortening the procedure of calling experts sequentially with examination, cross-examination and re-examination of each in turn. Moreover, the arbitrator was able to ask questions and follow up in sequence. If a court appoints a single expert, time is saved by adopting the expert's evidence instead of the sometimes lengthy process of identifying differences between expert's opinion and trying to resolve them. When one or both expert mistakes his role and adopts an adversarial position, acting as a hired gun, the time taken to resolve differences between experts can be very lengthy.

An alternative method taking far less time is mediation. Whereas hearings in arbitration or in court can last for many weeks or many months for complicated cases, mediation is usually conducted over a period of a few days, at the most, after preparation. A mediator does not make a decision. His role is to facilitate a meeting between the disputing parties and to encourage them to come to a mutually agreed-upon settlement. The two parties try to find a middle ground with the least pain on either side. Mediation saves a lot of time and expense: a mediation might last one or two days, whereas an arbitration hearing can last for weeks or months depending on the number of issues in dispute.

5.2 What can go wrong

Despite best endeavours, sometimes things do go wrong. A study of failures provides lessons to be learned. The first lesson, of course, is not to make the same mistake twice.

5.2.1 Lessons

For civil engineering works, including deep excavations, society does not want damage or fatalities. Fatalities can result in charges of manslaughter and custodial sentences. Damage should not occur, but if damage is inevitable then repairs or replacements must carried out and paid for.

Why devote a whole Section of this book to failures and collapses? Not because engineers are irresponsible, and failures do not happen frequently. The infrequent failures that occur are valuable lessons from which all engineers should learn.

Civil engineers belong to a learned society and a professional practice. The learned society seeks to understand how things behave. How strong are building materials such as steel and concrete? How much loading can they withstand and remain elastic? When loaded, do they deform, and when unloaded do they go back to the size and shape they were before? How ductile are they, and if they yield, do they absorb energy like a steel air-raid shelter, or do they crush like eggshells? Intelligence is gained from experience, and if materials are tested and if their properties are well understood then they can be used reliably. Professional practice requires engineers to act responsibly and to respect society.

Engineering in general can be said to create things for the benefit of society. In factories, manufacturing makes use of materials that have been produced to meet a specification within specified and verified tolerances. Civil engineers, especially geotechnical engineers, have a more difficult task than manufacturing by working with natural materials that are inherently variable. As I recall, a checking engineer once said to me something along the lines of "*I am concerned about the risks of what you propose, and I consider that you are interfering with Mother Nature*" (I paraphrase). My response was "It is the art and the practice of civil engineers to work with respect for the natural environment to create useful facilities for the benefit of mankind."

Not only do civil engineers have to deal with variable ground; in areas of seismicity, civil engineers cope with the effects of enormous releases of energy during earthquakes. Natural phenomena are uncertain. For earthquakes or tsunami, one does not know when the "big one" is coming or how big it will be. For geotechnical engineers there is always uncertainty. What is the ground like between these two boreholes? There is sand in this borehole; is it a trapped lens or an extensive aquifer? Nobody can write a specification for the properties of the ground. The ground has to be explored; the soils have to be identified and tested. Investigation and testing are matters of sampling aimed at understanding the range of properties and their more frequent values. Furthermore, the properties of the ground can change with time and can change as the result of the engineering works. For example, a deep excavation for an eight-car underground railway station can involve removing nearly 200,000 tonnes of soil. As a consequence, the ground below the excavation is unloaded and can swell and the ground below formation level can weaken. The strength that is measured in the ground before excavation is not necessarily going to apply as the excavation approaches the final level.

In order to manage uncertainty, civil engineering makes use of several layers of safety. First, there are factors of safety applied to the uncertainty about the nature and properties of the ground and to the expected loading. Second, designs are checked. Third, the ground conditions that are exposed during construction are inspected and tested in order to verify the assumptions that were made during the design stage about the ground conditions. Fourth, the works are monitored during construction including measuring deformation to verify the expected performance. If any disparity is encountered, remedial works are considered and implemented as necessary. For example, when the strength of the ground is relied upon for support, the design considers not only the typical strength that can be expected to occur most of the time but also the weakest possible case. Specifically, for limit state design, the Eurocodes specify that for serviceability the strength that shall be considered to apply for most of the time should be a moderately conservative value above which most of the measured strengths are found; a partial factor is applied to the moderately conservative value for strength. Factors also apply to loading. What loading is expected and what extreme events could happen? There must be a margin of safety. Some 40 years ago, FoSs were not defined and, say for slope stability, engineers were expected to responsibly determine the reliability of their evaluation of the strength of the soil and the consequences of a failure and to propose an overall FoS according to the level of risk. The overall FoS is the ratio of the restoring forces that could be obtained if the strength of the soil is fully mobilised divided by the disturbing forces due to gravity on the mass of soil that is above the assumed worst potential sliding surface. Slope stability analysis was based on assuming a potential sliding surface whereby soil above the surface would fail by sliding on the surface. The disturbing force was computed from the self-weight of the soil acting down the potential sliding surface, and the resisting force was calculated from the shear strength of soil on the sliding surface. The FoS is the ratio of the resisting force divided by the disturbing force. Values of FoS that were adopted were typically 1.5 for high-risk slopes and 1.2 for low-risk slopes. An overall FoS of 1.5 means that the computed restoring forces are 50% greater than the computed disturbing forces.

Some of the early calculations for stability of slopes considered plane cross-sections and adopted circular potential sliding surfaces. Sliding on a circular surface is a solid body rotation about the centre of the circle; disturbing effect and restoring effects were computed as moments of rotation about the centre. Later methods of analysis were devised to consider non-circular surfaces; modern computer continuum programs predict generalised slip surfaces such that the mass of soil above the surface deforms as it slides on the slip surface.

In 1978, as the result of multiple fatality landslips in 1972 and again in 1976, the HKSAR Government established the Geotechnical Control Office (GCO) in order primarily to prevent further fatal landslips and to set standards for slope safety. GCO produced a Geotechnical Manual for Slopes [23], which set out standards of good practice to be adopted for design of new slopes and for evaluation of the stability of existing slopes for engineers. The manual includes overall FoSs to be adopted for slopes with different levels of risk. This approach is recommended for cutting slopes that would be used for deep excavations, although Hong Kong is so heavily congested that few deep excavations have cutting slopes. The reliability of the assumed strength is an important factor to consider when assessing risk. No guidance is given as to the amount of ground investigation that is assumed when adopting these standards. No doubt, the guidance in the manual was based on the assumption that ground investigation would be carried out to the common standard adopted in the industry at the time.

For earth retaining walls, FoSs have been recommended for many years for sliding, bearing capacity of the foundation and global stability with potential sliding surfaces passing beneath the wall similar to a slope stability assessment. Calculations were adopted with un-factored loads and un-factored strength, and an overall FoS was adopted. As discussed in Section 3.10, the recent of limit state design as set out in Eurocode 7 [15] applies partial factors to the material strength and to the loading. Also, two states are considered. The serviceability state is intended to achieve satisfactory performance under expected range of conditions, and the ultimate limit state is intended to ensure no collapse under extreme conditions with a low probability of occurrence.

Codes of practice are drafted and regularly reviewed by panels of engineers experienced in the type of work. The purposes of the reviews are to ensure that the recommendations are indeed resulting in satisfactory performance and safety, taking into account changes in industry practice.

When checking a design, a frequently asked question is "*Has it been done before?*" Although "engineer" and "ingenuity" have the same root in Latin language, new inventions brought about by ingenuity can be risky. Prior experience is very valuable in appreciating how things work and what needs to be put into place to ensure that things do not go wrong. In the past, the engineering profession has benefitted from lessons learned. Some examples when things did go wrong and the lessons learned and consequential changes that were adopted by the profession are discussed below. The profession is not averse to invention and new methods; techniques are studied and simulated by numerical modelling or physical modelling. For example, in Japan many years ago regulations were adopted whereby earthworks for large dams had to be modelled physically. In response, many companies built large centrifuges to model earth-filled dams at increased gravities to achieve the appropriate scaling effects for gravitational forces in small-scale models using the proposed filling materials. The test in the industry is if something has not been done before, have sensible steps been taken to understand how it will work, and are measures in place to verify the assumptions?

During construction, verification of geotechnical design assumptions on site is a responsibility of the Resident Site Staff (RSS). For this purpose, the basis of the design and the geotechnical assumptions should be supplied by the designer to the RSS. Sometimes engineers from the design team are seconded to the RSS and are present on site for this purpose. Sometimes samples of soil are retrieved during excavation and can be tested in order to verify their properties. During construction, instrumentation is used for monitoring the performance of the construction works. As described in Section 3.4, AAA limits that are set by the designer are the basis of assessment of the performance.

Civil engineering is often epitomised as adopting a belt and braces approach, with redundancy such that if "*one breaks the other will hold and if both break my gaskins (trousers) will fall.*" Redundancy is such that a failure of one element of a system would affect only a part of the whole system and not lead to total collapse of the system. An important part of the design process is to ensure enough reserve capacity in the system that if a component fails it does not lead to a total collapse. If a component fails, then the load that it was carrying needs to be shed to the other parts of the system. Generally, when there was a total collapse two or more successive failures led to the ultimate total collapse. Deep excavations, along with many other civil engineering projects, are built by people with mud on their boots. The thinking of the industry is *"make sure that it is safe; and keep it simple."* Unintentional things can happen on site. A deep excavation site generally has a lot of heavy plant moving around. Heavy loads of spoil are regularly lifted on sites. Accidents can happen such as plant hitting one of the many temporary columns or struts, over-excavation or excessive softening of the formation giving inadequate support to the earth-retaining walls. There is usually a large workforce, mostly of men with only basic skills. They need to be properly supervised and regularly reminded of personal safety that, in turn, can affect the safety of the works. If everybody does their job responsibly, things go right.

5.2.2 Classical collapses

In ancient times, bridge builders were called "architects," literally people who build arches. Then "ingenuity" was used as a root for the word "engineer," which meant "one who brings about something." One can engineer an event. In another usage, "engineer" is one who drives or services an engine; that meaning is not adopted in this book. Famous civil engineers, such as Thomas Telford [24] and Isambard Kingdom Brunel [25] excelled in ingenuity and invented many early forms of civil engineering. There were many successes such as the Clifton Suspension Bridge at Bristol and the Tamar Bridge at Plymouth, both brought about by Brunel. On the other hand, there have been failures. Some of them have been classical collapses, from which engineers have learned. This section briefly describes some well-documented collapses of bridges because they capture the imagination and can be readily appreciated from photographic evidence. Lessons learned from these collapses apply to the civil engineering generally. It is also appropriate that engineers working on any type of civil engineering understand other branches of their profession.

A common feature of major collapses is the multiplicity of small failures leading up to the final collapse. More than 100 years ago, when the engineering profession was in the hands of a few, some fundamental lessons were learned. Some notable failures resulted in a fundamental increase in understanding of physical phenomena and the properties of new materials. In the U.K., the Industrial Revolution of the 1700's brought about the need for extensive infrastructure including canals, railways and ports, which civil engineers planned and built. In addition, old materials such as cast iron were produced in ever-increasing quantities, and new material such as steel were being created and put into use. Notably in civil engineering, failures of bridges have been historical landmarks. For example, in 1871 construction began on a long railway bridge crossing the River Tay in Scotland with many piers and spans built from brickwork, cast iron and wrought iron. Immediately after its construction, there was a short history of minor failures. Cast iron components were cracking and braces were working loose. In 1879 during a violent storm as a train crossed the bridge, the bridge collapsed. The train and its passengers fell into the river. There were no survivors; only 46 bodies were recovered from about 75 people on board [26]. The principal reason cited for the failure was lack of consideration of forces from the wind; however, many issues were identified, including poor casting of cast iron components, steel cottar pins falling out and so on. Subsequent to the failure, the Board of Trade set up a commission that led to regulations, whereby all bridges had to be designed for a pressure of up to 56 pounds per square foot, 2.7kPa, depending on the wind speed for the location.

A second famous collapse giving insight into the effects of wind occurred more than 60 years later. In 1938, work began on the Tacoma Narrows Bridge, in the U.S.A [27]. At the time, this was the third longest span bridge in the world and the first suspension bridge in the U.S.A. When the bridge was completed and put into use, it was observed that vertical motions (vibrations) of the deck were induced by strong wind. The effects were studied, and attempts were made to control the vibrations. On 7 November 1940, the wind-induced vibration increased dramatically, and the bridge collapsed.

At first it was thought that failure of the Tacoma Narrows Bridge was resonance due to vortex shedding because it was thought that the Kármán vortex street frequency was the same as the torsional natural vibration frequency of the suspended deck. This was later found to be incorrect. The actual failure was due to aerodynamic flutter [28], which is now well understood. All large structures, including bridges and buildings, are subjected to dynamic wind testing on models. Wind, being a natural phenomenon, is difficult to predict in terms of magnitude and frequency. Historical records are analysed statistically, atmospheric modelling is used to estimate magnitude-frequency relationships for given locations for exposure to wind on structures and physical and numerical modelling are conducted to determine the responses of structures to the forecast winds.

In addition to wind, marine structures are exposed to effects of waves. Forecasting magnitude and frequency of waves is similar to forecasting wind making use of historical data and numerical modelling. For effects of wind and effects of waves, relevant authorities define standards for design for serviceability events and for extreme events. Serviceability events, such as those with a return period of up to 1:100 years, are specified such that during and after an event the structures are required to remain serviceable. Extreme events, with lower probability of occurrence, such as 1:1,000 years return period, are specified under which conditions collapse of structures should not occur and the degree of damage should be limited in extent such that the structures can be brought back into a serviceable state. During my early days of professional training, I worked on the numerical modelling of wave forces on a proposed gravity oil production platform for use in the North Sea. In 135m of water, the maximum credible wave was 37m high, as I recall. There were no prescribed standards in those formative days, and insurance providers, worked closely with oil companies and civil engineering designers to evaluate risk in a comprehensive manner where probability of occurrence, initial fabrication and installation costs and time, consequential damage and cost of insurance were comprehensively evaluated. I played only a minor role in the design for the project, but the philosophy of risk-based design contrasted with the deterministic design that was in place at the time for deep excavations. It seemed that uncertainties of the ground with respect to type, strength and stiffness faced by a designer in choosing a typical value far exceeded the margin set for some of the factors in use at the time and that field verification of design assumptions was equally important. I considered that such design assumptions ought to be fully conveyed to the site supervisory staff to keep designers informed about site observations. For some projects the designers supplied the site supervisory staff, but my concerns did not sit well with the occasional practice of appointing site supervisors independently from designers. Shortly after learning about risk-based design, I went on to supervise the construction of a vehicular underpass in Birmingham, U.K., and was fortunate to meet designers who took pains to keep me informed of the basis of their design and what to watch out for on site.

After the collapse of the Tay Bridge, cast iron and wrought iron were superseded by stronger and more ductile steel as material for building bridges. An issue is the weight of the bridge itself. In the interest of designing less heavy bridges, bridge designers seek to make use of lighter weight and stronger materials. For example, reinforced concrete is stronger in tension and bending than masonry. Steel is stronger than cast iron and wrought iron. Many bridges were built of steel, reinforced concrete or pre-stressed concrete. Large-span bridges, for which the weight is very important, made extensive use of steel. Lattice steel girders were fabricated from rolled steel sections, but steel rolling mills and transportation limited the structural sizes. Larger size steel beams were fabricated by welding together steel plates to make plate girders. The first application of a steel box girder for bridges was for military use. The Martel Bridge, a modular box girder bridge that could be transported by an adapted military tank and was suitable for military applications, was adopted by the British Army in 1925 [29]. Box beams are considerably stiffer than the more traditional I-shaped beams made to the same structural depth. In principal, a bridge with steel box girders could be made lighter in order to carry the same loads as a plate girder bridge. The dynamic effects of wind, as are understood since the collapse at Tacoma Narrows in 1940, led designers to consider aerodynamic bridge decks. Aerodynamic shapes, like aeroplane wings, are enclosed and steel box girder bridge decks were a natural choice to be fabricated in aerodynamic forms. Considered state of the art at the time, the Bosphorous 15 July Martyrs Bridge was designed in the 1960s and was opened in 1973. It comprises a steel box girder deck with a central suspended span of 1,074m. However, at the same time, in 1970 and 1971, 51 people died due to the collapse of three other steel box girder bridges: the Westgate Bridge in Melbourne, Cleddau Bridge in Wales (known as Milford Haven Bridge at that time) and South Bridge Koblenz in Germany. All three bridges had decks fabricated from steel box girders.

The Westgate Bridge, Melbourne, Australia, was under construction in 1970. The main span of 336 m is a steel box girder cable-stayed bridge; approach spans to either side were supported by steel box girders. During construction, the steel box girder for the end span between piers 10 and 11 collapsed killing 35 construction workers. Also in 1970, a steel box girder bridge was under construction at Cleddau, then called Milford Haven, and part of it collapsed during erection. The collapse at Cleddau in Britain led to a committee of inquiry into the design and erection of steel box girder bridges; called the Merrison Committee [30], it concluded that the cause of the collapse of Milford Haven Bridge was the inadequacy of the design of a pier support diaphragm. The collapse of this important part of the structure resulted in the catastrophic collapse of the bridge. A Royal Commission was set up to investigate the failure of the Westgate Bridge [31]. The causes of this were much more complicated, primarily those of the adoption of a newly conceived and previously untried and inadequately checked method of erection and failures of site organisation and communications between principal parties. One of the causative factors was the buckling of "K-plates" that were temporary cross-bracing to the steel box structure.

The phenomenon of elastic instability that is the buckling of slender struts was recognised by a Swiss mathematician, Leonhard Euler, in the 1700's. However, the evaluation of the buckling of thin steel plates including stiffeners, such as were used for the diaphragm of the bridge at Cleddau, when subjected to not only axial loads as formulated by Euler but also subjected to bending and shear in the plane of the plates was not fully investigated before construction. Studies were carried out at the direction of the Merrison Committee and design for stiffened steel plates was put onto a rational basis. The Merrison Committee prepared interim design and workmanship rules [32]. These contained detailed and specific rules for the analysis of stress in box girders and for the design of steel plated components in complicated stress fields. The rules apply to steel plates with and without stiffeners and for their connections. The rules address the effects of residual stresses due to welding and fabrication tolerances. The implementation of the recommendations led to wide-ranging changes in contractual procedures and in checking designs and construction procedures for steel box girder bridges. Since then steel box girders built prior to the publication of these standards have been examined and upgraded where necessary; new standards have been adopted for adopted for new bridges. Since then major collapses of this type have not occurred.

I was fortunate in 1970 to assist Brian Richmond, a former Partner of G. Maunsell and Partners, checking the steel diaphragm for the Cleddau Bridge and his independent report in draft on the construction methods for the Westgate Bridge just before its collapse. I put this experience to use working for him on the design of diaphragms above the bearings for a steel box girder bridge at Temerloh in Malaya. I then did some computer modelling for Brian in conjunction with Messrs. Flint and Neil on studies of stiffened plates as proposed by the Merrison Committee. This work was the basis of the next code of practice for steel bridges [33].

As mentioned at the start of this section, one might well ask "What is the point of discussing major collapses of bridge decks in a book about deep excavations?" It is not morbid curiosity. It illustrates vividly how things have gone wrong in the past and how responsible engineers can learn from experience and work with safety in the forefront of their mind. Collapses can have far-reaching consequences and multiple fatalities. That is not all. Deep excavations are large and complicated involving several disciplines. Engineers who work on projects with multiple disciplines should understand not only how their discipline fits into the whole scheme, but also what the other disciplines are and the basics of their scope sufficiently to understand when input from another discipline is needed. For example, a geotechnical engineer who knows about properties of soil ought to appreciate that steel can buckle and that struts should not be braced laterally against a wall that moves and induces curvature in the very struts it is intended to brace. Certainly, team leaders and project directors should have more than a working knowledge on all related disciplines.

In the formative days of GEO in Hong Kong in the 1980s, a committee was appointed to draft a model training programme for geotechnical engineers to achieve professional membership. The focus was on specialisation. By contrast, I was administering an in-house training programme, modelled on the training for civil engineers, including not just experience of all aspects of geotechnical engineering but also basic knowledge of drafting contracts, structural engineering, contract supervision, measurement and resolution of disputes. Several years later, more broadly based training programmes were adopted for geotechnical engineers.

5.2.3 Collapses of deep excavations

Collapses of deep excavations and failures on site are important learning opportunities for engineers who work on deep excavations to appreciate what can go wrong and how to prevent such events. In this section, three examples of catastrophic collapses of deep excavations are discussed. The first example is not complicated, and the lesson learned was to adopt qualified supervision during construction. The second example is discussed at length because of the many things that went wrong. The catastrophic collapse followed a series of failures of individual components of the system. Some of these failures have been observed on other projects, but, on their own, they did not result in catastrophic collapse. Also, because some 15 years have elapsed after the second example there has been time to observe not only follow up actions that were adopted but changes that have taken place as a result. The third example followed four years after the second and bears some similarity with it indicating the need to disseminate lessons learned from failures.

5.2.3.1 Collapse of Queen's Road Central, Hong Kong 1981

In April 1981 in the heart of Hong Kong's financial district alongside Queen's Road, construction of Gloucester Tower was underway, including excavation for a two-level basement. The temporary retaining walls for the deep excavation alongside Queen's Road collapsed along with half of the three lanes of Queen's Road, as shown in Figure 5.3. As a result, about 45,000 telephone wires were severed in days when mobile phones were exclusively military equipment. The excavation and the connected

Figure 5.3 Collapse of Queen's Road, Hong Kong 1981.

basement to Edinburgh Tower were flooded. For most of the month of April, traffic was diverted. Queen's Road was handed over for urgent reinstatement to the contractor for the basement; the first activity was to attempt to recover some of the telephone connections. The space was partially occupied by the Hong Kong Telephone Company. The head offices of six banks were closed for four days until gas mains were assured to be secure. The telephone company recovered about 29,000 connections and the contractor for the basement, without hindrance from traffic, sped up construction, completing the basement within weeks, less time than envisaged before the collapse, and then reinstated the road.

The ELS for the basement was a temporary wall of steel sheet piles driven to below the final excavation level and braced by two levels of steel struts. During the driving of the steel sheet piles, bedrock was encountered alongside Queen's Road above the level for the final excavation. Excavation went ahead and two levels of steel struts were installed. In order to cast the base slab, the lower level row of struts was removed. The steel sheet piled wall then had only the top-level row of struts and no toe-in at excavation level. The walls were free to rotate, and since the soil pressure was greater below the struts than above the struts, the walls did rotate and the toes of the walls kicked in. The remaining struts and waling went out of alignment, and half the road collapsed. Nobody had changed the design or the method of working to account for the sheet piles not being driven to below excavation level.

The lessons learned led to changes at the time by the BA as follows:

- Site supervision is required by professional grade engineers.
- The Designer's senior engineer is required to visit the site regularly.
- As-built record plans of temporary walls for excavations are to be submitted before consent is given to excavate. Any changes in the toe levels shall be justified or the design for the bracing shall be changed to accommodate the changed levels if necessary.

Subsequent to registration of geotechnical engineers under the Buildings Ordinance, the requirements have been revised such that full-time supervision is required on site by a professional grade geotechnical engineer under the direction of the RGE, who is responsible for the project.

To date there has been no repeat of this type of failure of a deep excavation in Hong Kong.

5.2.3.2 Collapse of Nicholl Highway, Singapore 2004

In recent years, the most significant collapse of a deep excavation happened in Singapore in 2004 with four fatalities [34]. This collapse is a milestone in deep excavations because of the magnitude of the event, the numerous

causative factors that came to light, the lessons learned and put into practice and how engineering practice in Singapore has moved on. Since 2004, no
major collapses of deep excavations have happened in Singapore.

20 April 2004, in Singapore, a deep excavation alongside the Nicoll
Highway close to the Merdeka Bridge collapsed suddenly and completely.
The excavation for construction of a section of the underground railway
was 30m deep and about 17m wide. Prior to the collapse, the ground was
supported by temporary diaphragm retaining walls with nine levels of
steel strutting; excavation had advanced to a depth of 30m ready for the
tenth row of struts to be placed. About 80m length of the excavation collapsed. The steel strutting was distorted and displaced, and the diaphragm
walls collapsed, meeting each other in the middle of the excavation. Adjacent to the excavation, the ground subsided by as much as 13m and
included all six lanes of the adjacent Nicoll Highway to the north. Subsidence stretched back some 30m to 40m from the excavation into parkland
to the south. Fortunately, no traffic was involved, but unfortunately four
of the site personnel were killed.

Figure 5.4 shows the cross-section of the deep excavation. The site comprised about 3m of sand fill that had been placed as reclamation about 30
years before overlying about 33m depth of very soft marine clay, shown
in uniform grey shading in the figure. The reclamation was not drained,

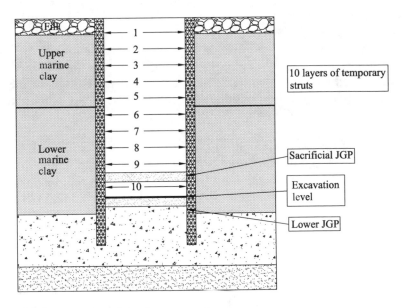

Figure 5.4 Cross-section of the deep excavation.

and site data showed some residual positive pore water pressure implying that the soil was slightly under-consolidated but had aged with creep for 30 years. Classical theory for consolidation identifies primary consolidation, whereby the imposition of a loading develops positive pore water pressures, which drive water out from the soil. The soil consolidates. When pore water pressures have dissipated, secondary consolidation, also called creep, takes place. This behaviour is an approximation developed from consolidation tests conducted on thin specimens of soil. In thin specimens, small residual pore water pressures were not taken into account. Soil does not learn theory. Creep takes place all the time, and after 30 years the site had experienced creep as well as the earlier rapid rates of dissipation or pore water pressure called primary consolidation. Notwithstanding slight under-consolidation and creep, the marine clay was a very soft clay with strength of about 5kPa at shallow depth increasing to about 20kPa towards the bottom of the stratum. Diaphragm walls had been constructed and keyed into the weathered weakly cemented sandstone called "Old Alluvium" underlying the marine clay. The excavation for a railway tunnel was planned to be about 33m deep at this location, and ten levels of temporary steel struts were proposed. At the ends of the steel struts, horizontal waling was used to transmit strut forces across wall panels. Some struts had splayed ends to spread the forces along the waling. The steel struts were required to be pre-loaded upon installation to reduce the lateral deflection of the diaphragm walls during excavation.

Pre-loading of struts is a common requirement. Pre-loading of struts is a good method to reduce deflections of the retaining walls. Without pre-loading, struts are compressed as they take up the load and the walls deflect forwards. A strut under a full load compresses about 0.1%, which is 1mm per metre. Thus, for a typical span for an underground railway station of 20m, steel struts compress by about 20mm. If this compression is taken up by pre-loading, the deflections of the two opposite walls are reduced by about 10mm each at that level. In some places pre-loading is a regulation. I consider that regulation of pre-loading unnecessary, and designers should determine the pre-loading they require; to improve fixity, designers should be allowed to adopt unloading for special cases. For example, unloading of struts located above a recently pre-loaded strut generates an increment in hogging bending moment in the braced wall at the level of the lowest strut. The increased hogging bending moment reduces the sagging bending moment lower down the wall, which forwards the deflection of the wall overall. This effect can occur naturally when adopting top-down construction, when an upper concreted slab shrinks and creeps and the lower level is pre-loaded. I have heard about only one other engineer who has recommended unloading struts in this way [35].

Because of the weak soils, large lateral deformations of the walls were expected, so two slabs of jet grout were constructed before excavation took place. They acted as buried struts to significantly reduce the lateral deflections of the walls. As described in Section 3.2.5, jet grout piles are about 1m to 1.5m diameter. Slabs are made by constructing jet grout piles overlapping each other in both directions, say 1.5m diameter at 1.3m spacing. To construct jet grout slab, the piles are only formed between selected depths within the thickness of the slab. For this excavation, two jet grout slabs were installed: a temporary one was just below the ninth level of struts and was intended to be removed after installation of the ninth level of struts; the second jet grout slab was installed below the final excavation level and was intended to remain permanently in place. This slab was intended to brace the diaphragm walls prior to and during excavation but also would resist base heave from the underlying weak marine clay.

It is commonly found that when structures fail, the structural members themselves are able to support the loading, but the connections are the weak points, and they fail. It is a case of "the devil is in the details." The struts for this project were steel beams with an "H" cross section (H-beams) that are often used for the purpose. The middle of the "H" is called the web, and the two other plates are called flanges. The horizontal struts were placed vertically with the H-section on its side so that the web was vertical, and the two flanges were located on the top and on the bottom, as shown in Figure 5.5. The waling also comprised H-beams but placed with the web located horizontally, like a letter "H" so that the strut was in contact with one flange of the waling as shown in the figure. The other flange of the waling was in contact with the diaphragm wall. A description of the connection and its significance in the events that took place leading to the collapse is given later in this section.

After the collapse, the area was stabilised by pouring in a lot of concrete. Some of the debris was taken away and the area was back-filled. This length of tunnel and the adjacent underground station were abandoned. At the time of the accident, construction of the adjacent three-level underground cross-over junction station at Nicoll Highway was well underway and the base slab already constructed. Shortly after the collapse, the station was back-filled and has never been used. The adjoining bored tunnels beneath the harbour had been constructed and lined and were awaiting follow-up track lying. They too were abandoned. A new alignment for the Circle Line alongside the abandoned one has been constructed. A new underground station and connecting tunnels have been built, and the line has been operating for several years. The physical loss to LTA was a junction station, two lengths of tunnels and the planned future connection.

The site was stabilised, as described above. The bodies of the four deceased workmen were not recovered. A short time later, the Nicoll Highway was reinstated. Because of the fatalities, the police interviewed 204 witnesses of fact and took conditioned witness statements from them. A Committee of Inquiry (COI) was appointed to determine the cause of

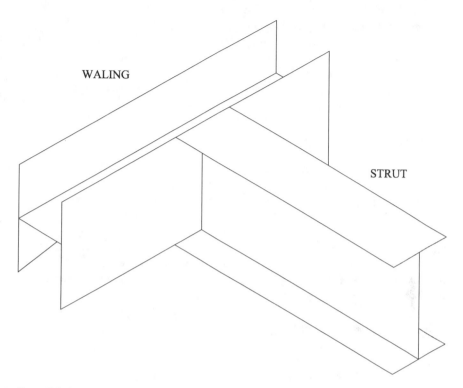

Figure 5.5 Strut to waling connection.

death and to make recommendations on actions to prevent such events in the future. The COI held hearings commencing August 2004; they published their final report in May 2005 [34]. A public inquiry was opened, and the Ministry of Manpower presented the witnesses of fact and introduced expert evidence. At the hearings, seven parties represented one to three experts each, with a total of 16 experts. Experts held meetings and produced an interim and final report in response to questions put by the COI. The COI then wrote a report on causation and identified liability and made recommendations. The LTA took out project insurance, and issues of financial settlement were discussed in camera after the publication of the report.

Prior to the inquiry, the first report in a local newspaper was *"Burst gas main causes collapse."* This story was soon disproven. The first lesson in this case is not to always believe what is reported in newspapers. Soon afterwards, the international press reported advice from experts that misuse of a computer program was the cause of the collapse. This story is still going around. Yes, the computer program was used in a way that underestimated the required capacity of the diaphragm walls, but the failure was initiated by the collapse of the bracing connection at the ninth level followed by progressive collapse of the system. There is an argument that if the bracing at the eighth level did not have the same detail for the connection, and if it could have withstood its intended design capacity, the overall collapse would not have happened. There was a requirement for redundancy, to check the design of the bracing for conditions of any one strut failing; however, when the strutting at level nine failed, the remainder of the ELS system could not withstand the redistribution of the loads and progressively collapsed.

At the inquiry, the sequence of events on site leading up to the collapse was educed. In February 2004, two months before the collapse, waling between struts and wall panels on the same excavation were found to have buckled, and the design of the connection between the struts and the waling was changed. Pieces of steel channel section were added across the top of the flanges of the waling to strengthen the connection. There are two weaknesses of this type of connection. One is bending the flange of a waling as shown in Figure 5.6 and the other is sway buckling of a waling as shown in Figure 5.7. At the Nicoll Highway, both mechanisms occurred.

Connections of this type between two H-beams are commonly used, and it is necessary to reinforce the connections with steel gusset plates to prevent these two mechanisms. Stiffeners are required to prevent these two mechanisms. A standard detail for this type of connection is shown in Figure 3.9. Stiffeners should be added, vertically spanning the flanges of the waling, above and below the web, in line with the vertical web of the strut. The stiffeners resist both modes of deformation. The stiffeners provide connections for the axial load in the strut for both the strut and

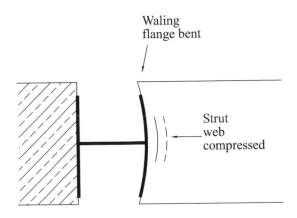

Figure 5.6 Flange yields.

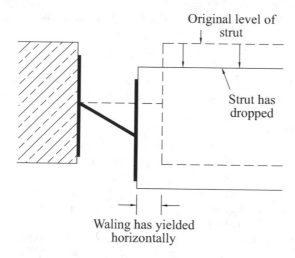

Figure 5.7 Sway buckling.

for the waling. By this means, deflection of the outer flange, as illustrated in Figure 5.6, is prevented. In addition, the stiffeners resist the sway buckling mode of deformation as shown in Figure 5.7.

On site a repair was attempted by welding a piece of U-channel steel across the flanges of the waling. Unfortunately, the piece of U-channel steel that was used on site did not provide adequate capacity. Subsequent to the collapse, laboratory tests were performed on copies of the unusual connection using new steel and careful welding. Mock-up connections were tested to failure; the tests showed that properly made connections of this unusual type had only about half the required capacity. Moreover, because there was no vertical stiffener in the waling, the connection was not very stiff in shear and failed with a sway buckling mode. Steel normally strain hardens and, after initial yield has occurred, marginally increases its strength; however, the unusual connection that was adopted exhibited sway buckling, as sketched in Figure 5.7: a brittle failure with massive reduction in capacity after first yielding. This type of connection was used for the seventh, eight, and ninth levels of strutting at the location of the collapse. The ninth level was the lowest level and at the time had the highest load; it failed first, and the ELS tried to shed additional load to the eighth level, which then also failed and total collapse followed. The collapse was a consequence of a detail that was not properly designed; the main struts had plenty of capacity. As if often the case, "The devil is in the details."

On the day of the collapse, 20 April 2004, excavating work had reached installing the tenth level of steel struts after installing and pre-loading the ninth level of struts. First thing in the morning, the strut to waling connection at the ninth level was observed to have buckled. Site staff planned to strengthen the connection. As it happened the location was fully instrumented with strain gauges on all the struts to measure the loads in the struts at every level. Inclinometers were installed to either side of the excavation and measured the lateral deflections of the wall on one side and the lateral deflections the ground immediately outside the wall on the other side, which amounted to almost the same deflections as the wall. Loads in the struts were monitored automatically every hour and could be read on a computer in the site offices. Before sending workers to strengthen the connections at the ninth level, the site engineer checked the monitored loads in the strut. All the readings at that location were less than the AAA limits, and he judged that it was safe to work at this location. He organised a crew to repair the connections for the struts at the ninth level. One might ask why the crew worked while something was failing. The monitoring data showed that the struts were loaded to well below their capacity. This and many other problematic issues became known during the inquiry and are addressed later in this section.

A full account of the inquiry and its findings are included in the final report of the COI [34]. The COI critically considered the circumstances leading up to the failure with respect to design standards, design of the temporary works, checking of the design, the sequence of construction, supervision, monitoring and control of safety of the works. The COI identified causes of the initial failure and the total collapse and cited errors and omissions leading to the collapse.

In summary, the collapse commenced with failure of the connections between the steel struts and the steel waling at the lowest level of strutting, the ninth level. The failure of the connections was brittle in that there was no residual capacity in the connections. The failure at the ninth level of strutting was followed by failure of the waling at the eighth level. Overall collapse resulted from the inability of the retaining system to withstand the redistribution of the loads arising from the failure of the strutting at the ninth level.

The COI found that the design of the connection between the struts and the waling involved two errors: an incorrect application of the design method to determine the capacity of the connection and the assumption that two splays would distribute the load from each strut; however, two splays were not used for all of the connections. The COI identified another fundamental fault: a wrong method was used in the computer analysis for the design of the temporary works. Next, the COI identified nine major contributory factors as follows:

- Inconsistencies between design criteria and codes
- Insufficient toe embedment for hydraulic cut-off

- Special geometry not taken into account
- Cable crossing disrupted the diaphragm walls and JGP slab
- Inadequate appreciation of complex ground conditions
- Inappropriate choice of permeability for OA
- Delay due to cutting walls for tunnel shaft
- Large spans left un-strutted for a long time
- Loss of pre-load in struts at levels 8 and 9

In addition, the COI listed the following contributory factors:

- No check of one strut failure in back-analysis
- Work did not stop in face of warnings on site
- Failure to implement risk assessment
- No independent design review
- Weakness in the management of construction changes
- Instrumentation system not effectively used

Civil engineering depends not only on proper design with FoSs, or load factors and material factors for limit state design. It also depends on checks and control during construction. Failures of individual elements of a system seldom result in catastrophic collapse. A collapse generally requires failure of more than one element or a series of related failures. In this case, the catastrophic collapse resulted from failure of some structural elements, namely a series of connections between the struts and the waling and a whole series of associated factors as listed above.

The COI made a series of recommendations to put right what had gone amiss on this project. The recommendations regarding engineering practice are paraphrased as follows.

- Temporary works were designed to reduced FoSs compared to permanent works. For major temporary works such as these, the design should adopt the same standards used for permanent works.
- The design of temporary works was checked in house. The design should be checked by competent persons such as an independent body and certified by a qualified engineer.
- There should be an over-riding ethic of safety and of ownership and management of risk.
- A specific code of practice for deep excavations should be prepared.
- Generally, numerical analysis or modelling should not be relied upon too much. It can only be used to supplement and not supplant sound engineering practice and judgement.
- Sophisticated computer software should be used only under guidance from specialists.

- Computer analysis must be well undertaken by competent persons. Those who perform geotechnical numerical analysis must have a fundamental knowledge of soil mechanics principles and a clear understanding of numerical modelling and its limitations.

In May 2005, BCA promulgated guidance notes to concerned parties [36]. The two major changes are that temporary works shall be designed to the same standards as permanent works and shall be checked independently. The areas covered by the notes include adequacy of site investigation, codes, standards, extent, groundwater conditions and existing building conditions. Issues of design include guidance on FoSs, soil parameters, water pressures, robustness, numerical modelling, sensitivity analysis and jet grouted piles. Issues during construction include multi-tier level monitoring, design review, independent checks, site inspection and approval, and instrumentation and monitoring.

In 2005 the LTA, the authority who operate the MRT, adopted similar procedures more specifically geared towards underground railway construction. They include, in addition, details of global stability checks, the use of software and drained and undrained analysis, back analysis, as well as maximum allowable movements and assessment of buildings and utilities. LTA appointed consultants to perform independent reviews and checks for the completion of the Circle Line and procures instrumentation directly, not via the main contractor.

COI stated that total collapse followed initial failure because of *"The inability of the rest of the system to withstand the redistributed loads after failure of the 9th level struts."* This conclusion of the COI embraces many issues. The system in its broader sense includes such design standards as load factors adopted, design assumptions, geotechnical parameters, computer modelling, preparation of design and checking, quality control of construction including installation of jet grout piles and preloading of struts, control of sequence and timing of construction, installation and maintenance of instrumentation, monitoring, exceedance of Alarm Limits, remediation of defective or damaged items, keeping track of design changes, proper back-analysis and implementation of contingency work. This is set in a background of insufficient attention to quality control, adherence to procedures, risk management and an insufficient culture of safety. The COI found deficiencies in all of these aspects. One of the root causes of the failure of the system was, ironically, that there had not been a fatality on an LTA construction site for 20 years and complacency had set in. A truism of nature is that order becomes chaos. Chaos does not resume order naturally. For example, snooker balls all over the table do not of their own volition move into a neat compact triangle. There are many ways of saying "Things go wrong." Diligence is always required in order to prevent things from going wrong. The story of this collapse could be told many times in the hope that people do not become complacent.

One should be cautious about giving an opinion before forensic evidence is available. Before the opening of the inquiry, the COI was given three experts' reports on the collapse. All three experts addressed the use of PLAXIS, and they used PLAXIS to analyse the collapse. The experts concluded that the collapse was due to base failure of soft clay at the bottom of the walls. Evidently, they were not aware of the inadequacy of the connections between struts and waling, but that was not all. Deep excavations are instrumented and monitored. Inclinometer instrumentation comprises vertical tubes that can be cast in diaphragm walls or grouted in the ground. They are surveyed by lowering a probe that takes readings going down and back up again. The probes are very sensitive and can measure the lateral deflections of the tubes, and hence deflections all the way down the walls. The vertical profiles are measured regularly and recorded. Profiles taken at different dates show the lateral displacement between those dates. Inclinometers are commonly used for deep excavations. An instrumented cross-section happened to be located right in the middle of the collapsed zone of the subject deep excavation. Inclinometers were located on either side of the excavation, and readings were taken about twice a week. The last reading was taken at midday on the day of the collapse when men on site were working to repair the buckled waling. The readings at that time showed inwards movement at the levels of the strutting but no inwards movement at and below the bottom of the wall. This shows clearly that base failure did not occur. Bulging of the walls higher up was consistent with inadequate strut forces and buckling of waling. The experts predicted excessive forwards movement at and below the bottom of the wall, which was not correct. Clearly some of the assumptions made by the experts in their analyses were not right. However, the real failure of the experts was not having asked to see data from inclinometers before reaching conclusions that were clearly in conflict with the data from site.

Immediately after the collapse, the international press released the story "from an informed source" that the collapse was due to incorrect use of a computer program. For several years that story went around and around. A few people remember it to this day. The original design calculations were developed using a computer program, Kasetsu-5X, which incorporates a refined beam on springs numerical model. Sometime before the collapse, unfamiliar with Katsetsu-5X, LTA's engineers asked for a check using a different computer program. PLAXIS was chosen for comparison, and some analyses were carried out. A meeting of LTA staff, site staff and a specialist in the use of PLAXIS was held only a week before the collapse when it was said that the program had not been used correctly. Possibly, with this meeting in mind, somebody jumped to the conclusion that a design based on using PLAXIS incorrectly was the root cause of the collapse as quoted in the international press.

The PLAXIS program can be used for drained or undrained conditions. When using the undrained setting, the computer program allows either the input of undrained shear strengths directly, called Method B, or the input of effective stress strength parameters c' and φ', whereby the shear strength is calculated using the Mohr Coulomb model, called Method A. In the program, Method A is an approximation. The undrained setting means no change in the isotropic stress, p' as illustrated in Figure 5.8 for one-dimensionally consolidated clay. When Method A is used, the value of p' remains constant and, as the shear stress is increased the stresses follow the vertical line AB in the figure. The strength envelope is the line OB in the figure. The lines OB and AB meet at the point B, which is the strength computed by PLAXIS. The actual undrained strength, c_u, is plotted at the end of the curved line AC at point C. The undrained shear strength is over-estimated by PLAXIS using Method A. Depending on the type of soil, the overestimation is of the order of about 16% to about 24%. Before the collapse, the manual for PLAXIS recommended the use of Method A. The manual has since been changed. The sources of the stories in the press asserted that the over-estimate is 60%, but they were mistaken. That only happens for soil that has been consolidated isotropically; this only happens in a special test in a laboratory, not in the ground. The stress path AB is how a soil would behave if it were drained and subjected to constant isotropic stress, not if it were undrained. One should be wary of press articles from "an informed source."

Saying that the strength is over-estimated is not helpful unless the consequence is understood. For example, if the correct strength is 20kPa the

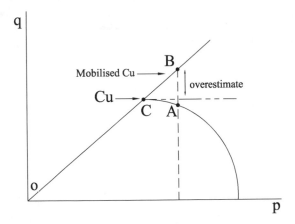

Figure 5.8 Overestimated strength.

incorrect estimate could be about 24kPa. What does this mean? For instance, at a depth of 30m using Method A the lateral stress on the outside of the diaphragm wall is only 2.5% less. This seemingly small change in applied load has a compounded effect on displacements, and the proper comparison would be to run the PLAXIS program using the two methods and to compare the differences. The outcome is that, for shallow depths of excavations, the use of Method A tends to over-predict deflections, but for greater depths Method A considerably under-predicts the deflections. When excavating 30m deep in soft soil, Method B predicts about twice the deflections compared to Method A.

It transpires that Method A was an inaccurate method. The method was only used as a check. The design was comparable to the results from using Method A, yet in the public arena the original design has not met the sustained level of criticism as the method that was used as a check.

There were more problems. The struts were required to be pre-loaded with jacks. At the instrumented location, strain gauges were fixed to struts before pre-loading. The gauges were read automatically every hour. On 5 April 2004, the strut at the ninth level was pre-loaded. The required pre-load was 6,100kN. The first reading of only 2,600kN dropped to 1,400kN within an hour. This major loss of pre-load is amazing but was not reported or investigated before the collapse. After pre-loading, the excavation was taken down reaching the tenth level on 17 April 2004. By 18 April 2004, the load in the strut increased marginally to 3,500kN, but it was a long way short of the designed load of 8,732kN. Nobody reported this large shortfall. Early on 21 April 2004 when the site engineer looked at the monitoring data, he would have seen a load of about 3,500kN, which was well below the AAA Limits. What he did not realise was that the measured force was only about 40% of the design load and was not providing the support to the wall that the designer intended. Had he looked at the inclinometer readings for the last few days he would have realised that the walls were moving inwards excessively at this level. The walls had exceeded the stop work limit on 23 February 2004, several weeks before the collapse. A back analysis, using PLAXIS, was accepted, even though it was flawed, and the AAA limits were raised. The new stop work limit was exceeded on 18 April 2019, only three days before collapse. Despite this, work was not stopped. There were only two weeks remaining before the excavation was due to be completed and handed over. Even when faced with liquidated damages a stop work limit should not be ignored. In the evidence from witnesses of fact, the only reference to monitoring of forces in struts was by the site engineer who on the morning of the collapse looked on line and saw that forces in the struts were less than the limits based on the capacity of the struts; he concluded that it was safe to work in the excavation.

The effects of loss of pre-load were to be seen at other struts going back several weeks before. Just prior to the collapse the total forces in all the struts was only two-thirds of the designed value. For individual earth retaining walls, the forces restraining a wall against sliding should be at least 1.5 times the forces that are pushing the wall forward. The total of the forces restraining this wall were measured to be two-thirds of the design value; this should have been a clear warning that the FoS, nominally 1.5, had probably been reduced to 1.0 or thereabouts and that failure by lack of support to the walls could have been imminent. It is unfortunate that there were no AAA limits for the minimum forces in struts necessary to provide safe support to the diaphragm walls. There were only AAA limits determined from the maximum allowable force in each level of struts. Had the minimum AAA limits been in place for the struts the site engineer might have been alerted to the fact that the forces in the struts were two thirds of what was required for the stability of the walls. Alternatively, had the back analysis in February 2004 been carried out correctly, the inadequacy of the strut forces might have been identified.

The COI identified many issues, and the final report and the summary of their report are worthwhile reading for anybody seeking a better understanding [34]. Further technical discussion of issues that went wrong can be found in my papers [37, 38].

The COI concluded that there was a failure of the system. What does this mean? The overall system of civil engineering is designed to avoid failures. The system requires factors of safety. It requires checking design. It requires monitoring the works, checking the monitoring, and back-analysis of the results. It requires response to action limits, and it requires stop work at stop work limits. The design is not an end to itself. Mistakes should be picked up by checking. Departure from the design assumptions, or mistakes whereby a design error is not picked up by the checker, should be reflected by the monitoring and picked up by those who review the monitoring. Provided everybody does their job properly, the construction is safe and successful and the overall system works.

Experience within the industry is that collapse of a physical structural system, such as an ELS system, usually involves more than one mistake. Usually with all the factors of safety and all the checks in place there has to be a sequence of at least three or more mistakes for a physical structural system to collapse. The structural system is designed such that even if individual components fail, the damage is contained, and total collapse does not ensue. However, no one involved should assume that the system is robust enough that inadequacy in his area of responsibility can be tolerated because of the reserve capacity in the system as a whole. Diligence is required, and complacency must be avoided. As regards management, considerations of safety should not be set aside in order to pursue a tight programme.

The BCA in Singapore took immediate action after the inquiry. They issued an Advisory Note [36] that promulgated many changes including the following:

- Deep excavations shall be designed to the same FoSs as permanent works with no reductions.
- Instrumentation shall be monitored independently of the main contractor.
- There shall be independent site supervision by a QP.

Amongst the lessons learned, it was also apparent at the time that for different stages of the excavation the forces in struts are expected to be different and engineers should specify different AAA limits for different stages of projects. Also prior to the collapse, engineers were focused on the load-carrying capacity of struts and set AAA limits to limit the maximum forces in struts. They had not appreciated that the purpose of the struts is to provide enough support to the retaining walls and that AAA limits should include minimum forces in struts for protection of the retaining walls.

5.2.3.3 A similar case

Did the industry learn from the collapse of the Nicoll Highway? No: only four years later in 2008, in another city in South East Asia, there was a collapse of an ELS system with several similarities to the ELS that collapsed alongside the Nicoll Highway. A deep excavation comprised concrete diaphragm walls, ten levels of steel strutting and a strut-to-waling connection that had close similarities to the connection used at the Nicoll Highway. However, instead of adding a piece of channel to the waling, they used a piece of steel angle section approximately half of the size of the piece of channel used at the site next to the Nicoll Highway. The capacity of the connection was not tested and its contribution to the collapse is not known. A fundamentally wrong detail should not have been replicated.

At this other site there were two other obvious problems. One problem was that the steel struts were quite long and required vertical and lateral bracing. Lateral bracing was connected to a side wall. It appears that the designer thought that the lateral bracing would be secure on the side wall. However, during excavation the side walls can be expected to deflect, and the lateral bracing therefore would be deflected and would deflect the main struts, thereby reducing their capacity instead of keeping them straight and maintaining their capacity. The second problem affected three sides of the remainder of the excavation that had diaphragm walls held by tie-back anchors cemented into surrounding soil. Monitoring results showed that the walls were moving forwards. That was to be expected, but they were moving at a progressively faster rate. When walls move forwards, secure tie-backs should become stretched and resist the movement with a proportionally increased force. These did not. The forces in the anchors were reducing, indicating that the anchors were yielding. A simple check showed that the anchors were located too close to the wall and that the ELS system of anchors and wall was starting to fail. Fortunately, stabilising measures were implemented in time and there was no collapse in those areas.

5.2.4 Ground water control

Lack of control of groundwater can cause big problems in deep excavations. In most cases, control of groundwater is best achieved by hydraulic cut-offs provided by deep walls. However deep walls can be expensive and grouting the ground to make it less permeable can be a cost-effective alternative. If ground water is already flowing, casting concrete for diaphragm walls can be a problem. Flowing water can displace concrete or it can leach the cement and fine materials out of the concrete before it sets. Likewise, grouting of the ground can be ineffective if ground water is flowing. In the case of deep sewer tunnels in Hong Kong, insufficient grouting before mining the tunnels led to quite large inflows after mining, and post-grouting when the water is flowing into the tunnel is wasteful since a lot of grout material gets washed into the tunnel and is not effective. Stage 2 sewer deep tunnels in Hong Kong adopted pre-excavation grouting [39] with testing to ensure that the limit on inflow would be achieved; this was found to be far more effective than grouting after excavation trying to stop water that was flowing into the tunnel. Recently a deep access shaft for tunnelling in Singapore required grouting to keep out the water. When rock was reached below the diaphragm walls, more grouting was needed because water under a lot of pressure at that depth was jetting into the shaft. Much of the grouting was ineffective because it was washed into the shaft. The outcome was not a collapse, but a considerable amount of time and materials were expended doing repeated remedial grouting with claims for time and costs.

Dealing with karst (cavity bearing limestone) is an extreme groundwater problem because cavities can be very large, and water can flow freely through the cavities in some cases. Recently, at a site for a deep excavation overlying karst in Kula Lumper, the water table in the karst was found to be lower than the perched groundwater table on top of overlying fine-grained soils. Consequently, a ground investigation borehole lost its flushing water when the karst was reached. The outflow of water into the karst dislodged some of the soil, and a sinkhole appeared close to the borehole. This gave a warning that slurry could be lost from diaphragm wall trenches into karst in a similar way; the karst had to be sealed before constructing diaphragm walls. Karst can have large cavities and conductive passageways such that large volumes of grout can just flow away. Volumes of grout can be reduced by using quick-setting grout to block the conductive passageways or parts of the cavities. Costs of grout can be reduced by adding fillers such as sand or sawdust to the grout.

When excavating below the water table in karst, there is a big risk of water flowing up through the base of the whole excavation whereas in less conductive ground the flow is mostly underneath the walls at the perimeter of the site. Therefore, in karst it can be necessary to provide a hydraulic cut-off not only below the perimeter walls but also across the whole of the base of the excavation and not just under the walls.

5.2.5 Protection of adjacent structures

Deep excavations in developed locations must protect adjacent land, utilities and property. Various examples have been referenced above and they are described more completely here.

5.2.5.1 Supreme Court, Hong Kong

In the 1970s, during construction of the first line for the underground railway in Hong Kong, the excavation for Central Station was next to the Supreme Court Building, which is now the Legislative Council building and a declared monument in Hong Kong. The building dates from 1910; it is in neo-classical style with ionic columns and constructed from masonry faced with granite blocks. The building was erected on reclaimed land. Its foundation was formed by driving hundreds of Chinese fire tree trunks into the mixture of reclamation materials and silt on the site. Consequently, the building is in effect floating on a timber raft. Such a foundation system requires the level to be maintained at a constant, and a groundwater replenishment system is installed to replace groundwater as required [40]. While excavating for the diaphragm, walling boulders of strong granite were encountered and heavy chisels were dropped repeatedly onto the boulders in order to break them for removal. As work progressed, the building settled and moved forwards towards the diaphragm walling. Masonry clad in large blocks of granite is very brittle; cracks appear in the joints between the blocks, and the joints become wider as movement continues. Lateral spread of this type is more damaging than the subsidence for an old masonry building.

It is important to assess the type of construction for nearby buildings and their condition since these two factors affect the amount of ground movement that they can withstand without damage.

5.2.5.2 Mong Kok

Construction of the first underground railway line in Hong Kong commenced in 1976 and was generally a success. Because of the very high population density that generated patronage along the route, the new line paid for itself in eight years, giving confidence for further lines to be built in quite quick succession. At that time, one of the most densely populated locations along the line, and in the world, was Mong Kok (Cantonese for "busy place") with more than 10,000 people per hectare. The density is one person per square metre. Before it was developed, Mong Kok was a low-lying marshy area, but by 1976 it had many buildings of various heights up to 17 floors and various types of foundations. The streets were quite narrow, and the underground railway stations were constructed very close to buildings. Some ground water seeped into the deep excavations, but the former marshland was very compressible, and settlement resulted. There was a problem, not because of the magnitude of the settlement, but because buildings on different types of foundations taken to different depths experienced different amounts of settlement, which created trouble at the joints between buildings. In response, the BA issued a practice note on dewatering in foundation and basement excavation works [22] to prevent damage due to dewatering. A local rule of thumb is not to lower the water table by more than one metre in the vicinity of buildings generally in Hong Kong. Since then, there has not been an issue with buildings settling differently due to lowering of the ground water table during deep excavations, although there have been issues with lowering groundwater by more than two metres due to tunnelling.

5.2.5.3 King's Road, Hong Kong

The third underground railway line to be built in Hong Kong was the Island Line, which runs along the north side of Hong Kong Island passing through many intensively developed areas. At the time, King's Road, North Point to Shau Kei Wan, was flanked by almost continuous mixed development of typically six-storey old buildings and contemporary high-rise buildings. As-built plans for private development have been lodged with the BA for many years and engineers can gain access to, and copies of, old plans. Importantly, when assessing the effects of dewatering in compliance with practice notes, copies of foundation records and framing plans for buildings are referenced. Records of piles founded on rock are also used when assembling a geological model for the vicinity, which usually shows types of rock and rock head. Whether the ground at tunnel level is rock, soil or a mixture at rock head is important information for tunnelling engineers. As-built foundation plans for a certain building showed piles driven to rock head. With roads to either side, other buildings were not immediately adjacent to the subject building, and the rock head along the north shore of Hong Kong Island is very irregular due to steep slopes and tropical weathering. The rock head profile along the line of some tunnels went up and down, but that was to be expected. The tunnelling was in progress; when it was close to the subject building, rock was not encountered. Some of the rock slopes between North Point and Shau Kei Wan were precipitous, and as the tunnels were to the north side of the buildings, they were expected to be downslope from them so that the tunnelling in soil was not a surprise. It was a surprise when the buildings settled suddenly and to such an extent that they were evacuated. Subsequent investigation identified that rock head beneath the buildings was quite a lot lower than shown on the as-built records for toes of the piles that were certified as being driven to rock. A similar problem came to light in Singapore recently for a small estate when as-built records for "Block X" showed piles and records for "Block Y" showed shallow footings. An investigation to verify the foundations identified the reverse, "Block X" has shallow footings and "Block Y" has piles.

The obvious conclusion from these two cases is not to trust old as-built records and to investigate the type of foundations for buildings that are within the influence zone for the project.

5.2.6 How much rock will be encountered when excavating for diaphragm walls?

Diaphragm walls are often adopted for deep excavations in soil, and the equipment for excavating for diaphragm wall trenches is designed to operate efficiently, quickly and economically in soil. The presence of rock in trenches can result in slow and expensive progress. When rock is encountered, extra plant might be needed to meet the contract completion date, and the site area might not have space enough to accommodate more plant. Cutting rock in a trench using a cutting machine might take ten times as long as excavating soil, and breaking rock by chiselling might take longer. Therefore, there is a need to identify and quantify the amount of rock material, boulders and bedrock to be excavated by diaphragm wall panels at an early stage. At the design stage, insufficient boreholes might have been carried out to investigate the ground. For example, they might be of the order of 20m or more apart. For construction, diaphragm wall panels are typically about 6m long and pre-drilling is regularly adopted in Hong Kong at about the middle of each panel to determine depths to rock and, depending on the design, how much depth of rock is to be excavated in order to reach the required founding level. Even with borings at 6m centres, the top of rock in tropically weathered locations such as Hong Kong and Singapore is very irregular. An example is given in Figure 5.10 of Geoguide 2 [3] which shows the descriptive logging inside a 1.6m diameter shaft where the top of granodiorite varies by 6.3m.

When planning and estimating the cost and time for excavating diaphragm wall trenches at the time of tendering for the work, it is usually not feasible to pre-drill every wall panel location at 6m centres and certainly not feasible to explore at 1.5 m spacing of boreholes. One needs to be able to reliably estimate the profile of rock head between boreholes. Because of such variations, geological profiles that are interpolated between widely spaced boreholes and shown on tender drawings are often annotated with a warning note that elevations of rock head may vary between elevations determined at the borehole locations. What variation is to be expected between boreholes might or might not be stated. A method of interpretation derived from fractal analysis that can be considered for this purpose [41]. The method is valid if the data points are from the same set. In other words, the method should be valid within a single geological structural domain. The technique is to compute the difference in elevation of rock head between pairs of boreholes and plot the difference versus the spacing of the boreholes. With 20 boreholes there are only 19 adjacent pairs. However, if one considers all possible combinations of pairs, at one spacing, two spacing, three spacing and so on, then the number of pairs increases from 19 to $2C_{20}$ (20×19/2=190) number of combinations, which is 190 points for 20 boreholes.

The question is whether the method can be used reliably. Data for a site that is underlain by weathered rock for widely spaced boreholes is shown in Figure 5.9 as a plot of difference in the elevation of rock head at the borehole locations versus the distance between pairs of boreholes. The solid line on Figure 5.9 shows that overall along the length of the site the changes in topography are small, the site is almost flat, and at 90% occurrence the differences in elevation of rock head are up to 20m overall and between adjacent panels, at 6m spacing, the expected difference in elevation of rock head is estimated to be up to 5m at 90% probability.

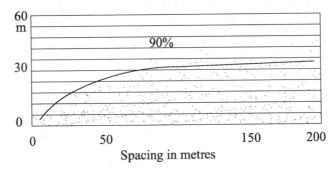

Figure 5.9 Borehole data before construction.

Pre-drilling was carried out at about 6m spacing. A plot of difference in elevation of rock head versus spacing of all boreholes and pre-drilling is shown in Figure 5.10. In this plot, differences in elevation of up to 45m are shown. The solid line is 90% probability for this data. The 90% probability line is similar to the pre-tender data shown in Figure 5.9, and the probability is similarly up to 5m difference in elevation at 6m spacing. This study suggest that the method appears to be useful provided the data is obtained within the same rock structural domain. This statistical approach was suggested to me by Dr. Andy Pickles [42].

5.2.7 Wrong use of computer programs

5.2.7.1 Wrong method

The wrong use of a computer program was a feature of the story of the collapse of the Nicoll Highway in Singapore. As a member of the COI noticed, computer programs are getting more user-friendly. The programs are complicated, and input of data and operational commands need to be clear. Some programs supply default values for some parameters to assist users. Although the intention is good, if the user fails to enter site-specific values for the parameters, the program will run and produce incorrect results. As somebody once said to me, "*If default values are not over-ridden, every deep basement is analysed as if it is in weak sedimentary soils irrespective of the presence of rock on site*" (I paraphrase). Some of the default parameters are not that obvious; in effect, specialists rather than every engineer in the office should be using the programs. The potential for mis-use of complicated computer programs can be highly dangerous because users might not be aware of what values of parameters are

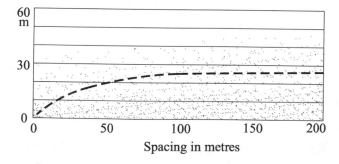

Difference in elevation

Figure 5.10 Closely spaced borings.

used yet can still obtain "results" with fancy contour plots from the output. Coloured plots can be very compelling, but they need to be realistic. I say this with the experience of my offices in Hong Kong holding over 30 licences for PLAXIS, three of them in 3-D. With, at times, over 500 staff, not everybody is using PLAXIS. We have a "PLAXIS Interest Group" and regular liaison with the PLAXIS agents and their help-line. The bottom line is that computers perform many calculations very quickly. Accurate software results in exactly computed results. Computers are powerful tools, but they need to be handled carefully. In Devonshire in the 1950s the farmers had a saying "*don't give cows muskets.*" Or perhaps more clearly, children should not play with fire.

5.2.7.2 Two drained analyses

A designer produced an undrained analysis that was cheap, a drained analysis that was expensive and a coupled consolidation analysis that modelled the seepage of groundwater and consolidation effects, which was an intermediate price. The contractor initially wanted to adopt the cheap undrained analysis but was prepared to adopt the design according to the coupled consolidation analysis. The checker felt that the fully drained conditions would apply. As it happened, two years later two experts examined the analyses and discovered that the coupled consolidation analysis showed almost 100% completion of drainage meaning that the ground was expected to be fully drained. they felt that the difference between the two analyses by the designer was not different degrees of drainage but was due to the two analyses having different assumptions about the wall. In one analysis, the wall was modelled numerically as a wall and retained water; in the other analysis, the wall was modelled as a beam on a slope, which would not retain any water. In this case, the designer had presented the coupled consolidation as a partially drained case and either had no understanding of the degree of consolidation that was achieved by his analysis or for some reason was playing dumb.

5.2.7.3 Negative settlement

A finite element programme was used to estimate the settlement of a foundation on a platform that had been created on a hill for an industrial facility. Parameters for the ground were assumed, and a settlement of about -100mm was estimated. The flaw was that the sign convention in the programme is negative upward. Why was the prediction that a foundation would rise by 100mm? At the start of the computations, the excavation to cut the hill down and form the platform was modelled to get initial stresses in the ground appropriate for the site just before the foundation was loaded. As a result of excavating the hill down to the elevation of the platform, the platform rose by about

110mm due to the removal of the top of the hill and showed a displacement of about -110mm. At this point, the displacements should have been reset to zero to measure the settlement from that stage onwards due to the imposition of the foundation. The net settlement of the foundation was only about 10mm, but because the initial rise of -110mm had not been zeroed, the total displacement shown by the computer was -100mm.

The lesson is that people who use computer programs should know what the program does, and in any case, there should be a check. If necessary, an independent check.

5.2.8 Wrong information

In an ancient game, a story is passed around a circle in a low voice. When the story returns to the first person, it is usually changed to the point of being unrecognisable. Reports from construction sites sometimes miss part of the story. Recently the term "virtual truth" has been coined and implies inaccuracy or a contradiction.

5.2.8.1 Not told everything

Giving respect to hierarchy is deeply imbued in several cultures in South East Asia and China. Several years ago, I saw a Senior Engineer who was very subservient towards his team leader, an Associate Director. It made me uncomfortable. A couple of years later, the Senior Engineer was rapidly promoted and was an Executive Director and senior to his previous team leader who was still an Associate Director. His attitude was completely reversed expecting his previous senior to be subservient to him. By contrast, at the time my Australian boss used to go to a nearby pub on a Friday night so that any of us, engineers of any grade and draughtsmen too, could find him and raise anything that we wanted to discuss with him. We could freely discuss any topic including open criticism of the boss to his face. He had the grace to treat us all as colleagues and not as vassals. As a result, we all had great respect for him. Several Asian cultures pay undue respect to seniors. Undue respect includes not asking one's boss what he wants but trying to second-guess what the boss wants and then to get praise if the guess was right and criticism if the guess was wrong. As for information about what has happened on site, site staff under management from office staff will err towards telling the boss what he or she thinks that the boss wants to hear. In the case of problems on site, site staff telling less than the whole story amounts to a form of cover-up. Some senior managers are consistently told stories that are less than true, but they believe the stories and then assert to others. In this way some practices do not achieve what is wanted, myths are perpetuated and even exaggerated with time.

5.2.8.2 One type of grout for every application

A case in point, misinformation that I encounter repeatedly concerns grouting. Grouting is a practice of pumping a liquid into the ground to fill voids or improve the ground. Grouting the ground can render the ground less permeable or strengthen or stiffen it. There is a range of grouting materials and grouting methods. Although it is commonly held that automobile manufacturer Henry Ford is reputed to have said *"You can have any colour car as long as it is black."* No one recipe for grouting will do everything. Around South East Asia, the mixture of cement and sodium silicate for grouting has been used for a range of purposes and in a range of soils has entered the realm of mythology. Sodium silicate is a solution of a salt in water, can permeate any type of soil that water can pass through and sets as a gel. It can reduce the permeability of ground quite significantly. The gel can weakly bind soil but does not greatly increase the strength of the ground. OPC has a particle size of about 0.04mm to 0.08mm. Grout that is a slurry of cement particles in water will penetrate smooth straight joints in rock that are wider than five particles (i.e. if the joints are wider than 0.3mm). For torturous paths through the pores between soil particles, the aperture of the voids between particles must be wider than about 1mm for cement grout to pass through. The minimum size of soil that can be permeated by cement particles is clean single-size sand. Cement slurry pumped into gravel can create a reasonable quality of concrete, perhaps up to 20MPa grade. However, cement that is pumped into soil with fine grains does not permeate the soil. Compaction grouting is when high pressure grouting is used to compress the surrounding soil and a bulb or lens of almost pure grout is surrounded by stiffened or strengthened ground. In some soils, high pressure may split the soil and further pumping widens and extends the split to form clacquages (veins of grout in slightly compressed ground). High pressure grouting, which forces cement into the ground as lenses or clacquages, does not materially alter the permeability of the ground. How many times I have heard that cement grouting was used successfully to control inflow of water on a job? I suspect that a lot of grouting has not materially affected the soil and that the good performance was because the ground did not need to be grouted in the first place.

5.2.9 Waterproofing

Many underground structures are built below the ground water table and waterproofing is required. Ordinary concrete has very low permeability it is not completely waterproof. The rate of seepage through good quality thick concrete is very small. In well-ventilated areas, the permeating dampness may not be noticed and might be tolerated. Seepage of ground water often brings dissolved salts through the concrete and can destroy the cement or corrode the steel reinforcement. These effects may not be noticeable for a long time after the completion of construction, and when the effects are noticeable inside the underground structure the damage can already be extensive and difficult to remedy. In order to achieve dry conditions, waterproofing is required by such methods as applying an external membrane or additives to the concrete. Problems can occur through poor workmanship in applying the waterproofing; this can be prevented by careful supervision and inspection. Joints between wall panels can be provided with an interlocking profile including a PVC water bar to prevent leakage through the joints. However, leakage through diaphragm wall panels happens from time to time (e.g. when concreting is done badly and there are inclusions of contaminated concrete, contaminated bentonite or simple included rubbish in bad cases). Site supervision and inspection are important. Most leakage of ground water is through joints between panels. The water bar can be damaged while excavating the adjacent panel; an accumulation of contaminated slurry can remain on the joint and create a water path. Leakages through joints can usually be identified during construction and remedial work undertaken. A leaking joint can often be blocked by grouting down the outside of the joint; resin grouting can be effective from the inside.

When constructed sequentially, each diaphragm wall panel is adjacent to a completed panel on one side and is open on the other side. At times, a closure panel is needed between two completed panels. If the space between completed panels is too narrow because previous panels are too large, or for some other reason, the walling excavator becomes stuck. If a stop-end from casting the previous panel gets stuck and obstructs the closure panel, the excavating machine cannot operate, and the closure panel is abandoned. A possible remedy is to construct one or two overlapping panels outside the intended alignment, if there is adequate space. However, the outside of diaphragm wall panels is generally very rough, and the overlap between panels is almost certain to leak. In dispersive soils, the incoming water can bring in soil resulting in settlement in the vicinity outside the wall. Sometimes a wide gap can result in a rapid inflow and rapid settlement. When an overlap panel is adopted, the overlapping joints should be sealed. Techniques such as grouting within the overlapped joints and blocking the outside of the joints with grout can be considered. It is prudent to make the closure panels longer than one pass of the excavating machine so that the panel is excavated by two overlapping passes; thus the risk of blockage is mitigated. Single pass closure panels should be avoided wherever possible. When relying on diaphragm walls to provide a cut-off to ground water it is prudent to test the watertightness of the completed wall by conducting a pumping test within the enclosure of the diaphragm walls before excavating.

5.3 What can go right

Civil engineering is mostly about success. For every publicised failure there must be thousands of projects that did not fail. There seems to be a fascination with hearing and repeating bad stories. Gossip is exaggerated. As Shakespeare scripted Mark Anthony to say, *"The evil that men do lives after them and the good is oft interred with their bones"* [43]. It is human nature to broadcast bad news such as collapses and to consider good news such as the completion of efficiently operating underground railway stations not newsworthy. It is hard to identify a deep excavation that is famous or even well known, except a small number of collapses. Efficient completion without notoriety is the objective of a good deep excavation project; it takes place and is finished unnoticed. Where viewing windows are provided in hoardings for deep basements in cities a few people have a look and perhaps marvel at some of the large and muffled equipment working away without affecting anybody.

The industry is geared up to carry out major civil engineering works safely and efficiently. The work is seemingly simple but complicated. In urban areas, there are many site restrictions such as lack of space, congested streets and nearby sensitive structures to protect. There can be several simultaneous activities on site involving several construction companies. Usually there are many instruments for monitoring the effects of the works and teams of staff monitoring the works and checking the performance. For example, a deep excavation for an underground terminus for a high-speed rail with 15 tracks deployed many thousands of people. The planning, design and construction of deep excavations is a great feat of civil engineering but lacks the fascination of more popular projects such as long span bridges.

5.3.1 Good engineering

To a layman, the definition of good engineering in deep basements is surely that nobody notices it apart from those who peered through the windows in the hoarding. Everybody takes for granted the permanent underground works, such as railway stations, carparks, shopping arcades, cinemas and ice-skating rinks. For engineers, the definition of a good project hinges on the question "so what was good about it?" "Within budget and ahead of time" is a cliché, but good engineers almost take completion on time and within budget for granted and do not boast about it. Commercially orientated people might ask how much profit the contractor made. Serious ones will ask what is new about this one.

5.3.1.1 MTR in 1975

If I look back to when I started working in 1970, they were formative years with a succession of innovative projects. Many of the innovations were generated by construction companies when tendering for projects seeking to win contracts. Checking requirements were less prescriptive than some are nowadays. In the early 1970s, there were not many basements deeper than three floors. Diaphragm walls were called "ICOS walls" because of the name of the Italian company, Impresa Costruzioni Opere Specializzate, who promoted the system commencing in 1938. Originally the excavating was done with grabs and resulted in 50cm and 80cm thick reinforced concrete walls. In 1975, I was tasked with producing tender designs for Paul Y Construction Co Ltd (Paul Y) for the Hong Kong MTR. The Chief Engineer for Paul Y was James Blake who later became Secretary for Works for the HKSAR Government. Paul Y had good connections with teams of men and women who hand excavated shafts called hand-dug caissons as described in Section 2.1.6. James Blake proposed that the hand-dug method be used to construct secant pile walls 150cm thick. He was cautious about using diaphragm walling for the project. He noted that there had been very little experience in Hong Kong with diaphragm walling equipment and that the required thickness of 150cm would require new large diaphragm wall equipment be made for which the rate of production was untried. He knew that the ground in Hong Kong often includes large relic boulders of strong rock, which would obstruct the use of grabs and require time-consuming chiselling, which would be a significant geological risk for the project. Since 14 stations were being tendered at the same time, he was concerned that the resources for diaphragm walls would be stretched.

There were advantages in considering hand-dug caissons. By employing over 700 workers, over 300 caissons could be constructed at the same time. If boulders were encountered, they could be drilled and split with steel wedges, which were commonly used for cutting back exposed rock on roadside cuttings in those days. James opted for the novel use of hand-dug caissons. His wisdom paid off. Paul Y was awarded contracts for Choi Hung Station and for Diamond Hill Station and used hand-dug caissons for the walls and for the plunge columns within the stations. Diamond Hill Station was intended to be widened in the future to become a junction station, so the longitudinal walls had to be easy to remove. The hand-dug caisson method of excavation was used for heavy steel king piles in alternate caissons and un-reinforced jack arches in the intervening caissons as described in Section 2.1.6. Other engineers in Hong Kong heard about hand-dug caisson walls and realised that the method required very little equipment and was ideal for constructing large diameter piles in steep hillside sites that are common in Hong Kong. Very soon there was a proliferation of hand-dug caisson foundations for heavy structures and hand-dug caissons for cantilevered retaining walls.

There were several other innovations for these two stations. Both were constructed by the top-down method, which was almost unheard of. As described in Section 2.2.2, the benefits of top-down construction included using permanent structure as bracing instead of steel struts or tie-backs. At Choi Hung Station one row of temporary struts was planned, but monitoring showed that the deflections of the diaphragm walls were less than anticipated, partly because of dewatering the ground outside the station. A limited amount of dewatering was proposed in the design; during construction it was found that there were no issues with settlement nearby and the amount of dewatering was increased beyond the designed level [44]. Dewatering reduces the total force on retaining walls; a back-analysis found that the struts were not needed. Accordingly, the row of struts was not used. There was a saving of cost and time in not providing and fixing the struts. In addition, instead of plant with reduced headroom beneath the struts, standard excavation plant with 5m headroom could be used and the final stage of excavation was completed more quickly than planned. The main reason for the removal of the struts was that some drawdown of ground water outside the site was permitted and resulted in less lateral earth pressure to be supported. The ground was relatively firm, and the nearby buildings were robust and tolerated a modest amount of settlement. The drawdown was planned by including sumps with filters in the bottom of some of the caissons before casting the walls; the amount of dewatering was controlled to limit the regional settlement with the benefit of lowering the lateral pressure on the walls. This was a practical example of adopting the Observational Method described in Section 3.5.

Because Diamond Hill Station was a standard two-track station, the roof was constructed as a flat slab close to ground level, and the top-down sequence of construction was adopted. Openings were provided for temporary access during construction. Choi Hung Station is unusually wide because it is a three-track station. In order to reduce the weight on the plunge columns only primary beams were constructed at first spanning between the diaphragm walls and passing over the plunge columns. A temporary gantry was provided running up and down the site, which could make use of the plentiful access to lift spoil between the primary beams until the concourse slab was cast and then hoisting was via the openings in the concourse slab. The gantry also served the production of pre-tensioned concrete Tee-beams. These were cast on top of the primary beams, which were the beds for pre-tensioning. The Tee-beams were turned around and placed longitudinally between the primary beams and topped with concrete, thereby completing the roof. At Diamond Hill Station, no gantry was built, but a technique was adopted that was fresh to Hong Kong: pumping concrete. A fixed pipeline was laid around the site to supply concrete on tap.

5.3.1.2 Preserving trees

From time to time, large and valuable trees are located on sites of deep excavation. Preserving large mature trees in this situation can be time-consuming and expensive. The technique that has been adopted requires encircling the trees with retaining walls at a large enough radius to enclose the majority of the roots of the tree. Then the ground beneath the tree and below most of the roots of the tree is mined and structural support erected beneath the retained ground. Temporary supports comprising steel beams are in place until permanent reinforced concrete structure is built to permanently retain the tree, the soil and its roots.

5.3.1.3 Redevelopment of a hotel

A very unusual deep excavation was part of the re-development of the Lee Gardens Hotel, Hong Kong. The excavation commenced with demolition of the building and the internal structure of the existing two-floor basement installing some propping to support the existing basement walls. The excavated material was the demolition debris. Additional bored piles were installed through the basement floor and a new basement and columns were constructed for the replacement high-rise building.

5.3.1.4 Deep excavation on a hillside

Another unusual deep excavation was for a private building at Number One May Road, Hong Kong. The site is unusual: alongside the site for Number One, May Road is a narrow road perched half way up a slope that is about 50m high and slopes overall at about 1 on 1 (i.e. 45 degrees to the horizontal). The original villa had been demolished and there was a landslip on the vacant site, so the site was profiled to a smooth cutting above May Road. There was concern about the stability of such a steep slope. Vertical boreholes sunk deep into the slope did not encounter rock. The proposed development of the site required two cuttings. The driveway from May Road was cut into the slope and climbed up to a top platform that was also cut into the slope. The top platform required an excavation to a depth of about 10m on the uphill side. Beneath the platform, a cutting for the access road was also about 10m high near the entrance. A photograph of the site during construction is reproduced as Figure 3.1. The site was considered un-buildable.

The project needed something out of the ordinary. Actually, there were three things out of the ordinary. The site was considered unsafe for building because a slope of soil 50m high inclined at about 45 to 50 degrees overall with ground water close to the surface at times is somewhat unprecedented and had not fallen down. In the 1970s, in an effort to improve the calculated low factor of safety for the overall slope and for the safety of the building located below the slope, horizontal drains had been installed at the toe of the slope to reduce the level of water inside the slope. Second, the site had been covered with chunam plaster and several projections of rock were taken to be corestones. Chunam plaster is made by mixing ten parts of soil and one part of cement and was commonly used on slopes to prevent infiltration of rain into the slopes. Chunam has been almost entirely superseded by sprayed concrete. James Kwong, a bright young engineering geologist noticed that the exposures of rock projecting through the chunam surfacing of the slope had joints that were very similarly orientated to those that were exposed in the stream beds to either side of the site. Additional boreholes revealed that for most of the site bedrock was close to the surface and that the previous boreholes had been located in a relatively narrow deeply weathered minor fault that traversed the site like a cleft in the rock filled with soil. To either side of the fault the bedrock was found to be exposed in places at the surface of the site, which gave the impression of corestones. The minor fault was found to be coincident with the horizontal drains that produced most of the water, whereas drains to either side had little or no flow. The notion of an unprecedented soil slope was dispelled, and the identification of a rock slope was confirmed. Hand-dug caissons were allowed in those days, and hand digging was an ideal method of construction for the two retaining walls for the excavations for the access road and the top platform. The walls that were 10m high had to be substantial. They had to be built before an access road could be built and before any heavy plant could be brought onto the site. Third, the design required analysis of tiered walls, the passive zone in front of the embedded portion of the top wall overlapped with the active zone behind the lower wall. It was decided to link the top of the lower wall to the bottom of the upper wall with reinforced concrete beams before the excavation in front of the lower wall. The design was completed in 1986 when computers had 286k processors and math co-processors had to be added in order to perform mathematical functions. The continuum computer program FLAC had been under trial in my offices for about six months beforehand, so it was used to model the slope stability, the effects of deep excavation in front of the hand-dug caisson walls and the interaction between the two walls. The designs were approved by the BA and were the first plans to make use of a continuum analysis for private building works in Hong Kong.

5.3.1.5 MTR Island Line

The first line, called the Modified Initial System (MIS) of the MTR in Hong Kong was opened in 1978. It passed through some of the most densely occupied locations of Hong Kong including Mong Kok with more than 10,000 people per hectare, one person per square metre at that time. The commercial success of the first line gave MTRCL confidence to build the Tsuen Wan Line and then the Island Line (ISL). With eight stations to be built, the ISL needed financing; a scheme was proposed to invite tenders for the development rights above the underground stations. Local development company Hang Lung bid a memorable HK$8 billion for the development rights. There was a hiatus when the property market had a big down-turn and Hang Lung exercised a right to not go ahead and the line was frozen for a while. Hong Kong had a habit of bouncing back, which it did, and the financing and associated developments went ahead. This financing model has been exported by MTRCL to other cities.

As part of MIS, only half of Admiralty Station was built with the two lines for MIS stacked on the north side. The construction of the ISL in the 1980s included the other half of Admiralty Station, which formed a cross-platform interchange station between the MIS; the ISL is now very busy. Additional lines have been planned and built. The Sha Tin to Central Link (SCL) is under construction connecting to Admiralty Station below the MIS. The recently constructed South Island Line (SIL) is below the level of the SCL. The SCL and SIL platforms at Admiralty are located to the east of the existing MIS station and pass beneath operating ISL tracks. Construction for the extension to Admiralty Station includes a deep excavation at the eastern end of the existing underground station structure through soil and into rock down to a depth of 45m below ground and covering a plan area of 3,550 m^2, large enough to house several escalators, plant rooms and pedestrian circulation spaces finally reaching the deep platform cavern for SIL. The first three floors of the excavation were constructed top down using substantial portions of the permanent reinforced concrete structure as lateral support and plunge columns as vertical support. Once bedrock was reached, earth lateral support was not needed and the rock was cut vertically with only rock bolts for support down to the bottom level. Then the remainder of the permanent structure was built from the bottom up. Excavating beneath the live tracks of the ISL required underpinning and mining, including blasting, which was carried out with the lines in service [45].

5.3.1.6 Multi-track deep station

In 1975, one underground railway station was considered to be a big contract. In recent years contracts have been let for three or four underground stations and related tunnels as one package. The recently completed underground West Kowloon Station in Hong Kong, which is the terminus for the high-speed rail from China, was built in a deep excavation 200m wide, 30m deep and 660m long [46]. This deep excavation was nearly ten times the volume of an eight-car underground railway station, which is typical for the MTR in Hong Kong. Once the "go ahead" was given in Beijing, the station had to be completed quickly. The need to excavate some 3.6 million cubic metres of soil and rock was facilitated by proximity to a sea wall and the prospect of marine transportation rather than contemplate road transportation through already-busy streets. Hoisting by cranes or gantries is relatively slow, whereas a haul road for trucks to drive down to the excavated level results in more rapid removal of spoil. Excavating a big pit with no bracing allows large plant to operate and match the haulage rates of lorries. The site had been formed recently as reclamation by filling over the seabed. The filling materials were considered permeable, requiring a hydraulic cut-off around the excavation. The underlying marine clay was found to be very soft and not yet fully consolidated under the filling materials requiring substantial ELS in order to complete the excavation to full depth to the perimeter of the site. Therefore, diaphragm walling was proposed around the perimeter of the site to provide both a hydraulic cut-off and, with bracing, earth lateral support.

During the bulk excavation, side slopes were intended to provide lateral support to the diaphragm walls. Steel strutting across the 200m-wide excavation was considered, but not adopted, because, having excavated in bulk to the final depth in the middle of the site bottom-up construction of the six underground floors could go ahead. By strutting off the completed portion of the permanent structure, the side slopes could be excavated, and the underground floors could be completed top down. The plan was sequential bottom-up and then top-down sequence. The side slopes to be excavated in under-consolidated very soft clay. With passive pressure from the embedded diaphragm walls, the side slopes would have to be very shallow. A stabilising system was devised using jet grout to replace some of the very soft clay to form steeper slopes, widen the bulk excavation, permit more bottom-up construction and reduce the top-down construction.

The main work was divided into two contracts for deep excavation and construction of underground structures. The contractor for the northern half adopted the planned scheme, whereas the contractor for the southern half decided to excavate and build entirely top down [47].

5.3.1.7 Hybrid sequence

It appears that the top-down method is often a contractor's design and bottom-up is more likely to be an engineer's design. Hybrid schemes for private building works constructing top down for the underground works and simultaneous bottom-up construction for the above ground works offer an early start on the aboveground works and potential time saving. The building works above ground can start once the ground floor has been completed to start the top-down construction for the basement and provide a working platform for both above and below ground work. Recently, the Sands casino and hotel in Macau was completed by this method, and I understand that six months were saved compared to the original construction programme.

5.3.1.8 Artificial islands

Sometimes extensive site preparation works are required before excavating can commence. One such location is the typhoon shelter, an area of moorings for small vessels; the other is the open sea between Hong Kong and Macao. The typhoon shelter is the location of a down ramp for a four-lane road tunnel constructed in soil connecting to short rock tunnels under Kellett Island. The open sea is the location for the Hong Kong to Zuhai and Macau crossing. This 20km route is mostly on a viaduct but a 3km section is below the sea in an immersed tube tunnel to mitigate the effects of ship impacts. The ramps from the tunnels to the viaduct were constructed by deep excavation. In both cases, the sites were created over water by building new islands. In the typhoon shelter, after temporarily relocating the moorings, a conventional sea wall was built from concrete blocks founded on a rubble mound. The sea wall enclosed an area for construction that was filled with sand to form an "artificial island." The island provided a platform from which diaphragm walls were constructed. The ground was excavated down to the formation level with bracing and a reinforced tunnel with a rectangular cross section was built bottom up. Filling materials were placed above the completed tunnels up to the elevation of the seabed. The diaphragm walls were cut down, the surplus sand and sea walls were taken away and the moorings were reinstated. For the sea crossing, in the open sea, two islands were built for permanent use, one at either end of the immersed tube tunnel. Permanent retaining walls were constructed by sinking 15m diameter steel casings and filling them with sand to act as sea walls. The space inside was filled with sand, which was then compacted, and diaphragm walling was used as retaining walls for construction of the ramp from the immersed tube tunnel up to ground level [48].

5.3.1.9 Deep excavation beneath a building

When the Airport Express Railway (AER) was built in Hong Kong, the terminus was built on some reclaimed land in front of the business district of Central. The AER station abutted the bus interchange beneath Exchange Square, a high-rise office building. The development rights above the AER station were sold, and the station was integrated with a large basement and foundations for the International Finance Centre, an 80-floor office building. Central Station of the MTR is on the other side of Exchange Square and across a very busy road. A 15m-wide subway was planned from the AER Station to the Central MTR Station. Because of the robust construction of Exchange Square, it was possible to excavate directly beneath the suspended base slab for the bus interchange at Exchange Square with an excavation that was about ten metres deep.

5.3.1.10 Deep excavation for foundations

Hong Kong has several buildings of 80 floors and more. In Central District, most of a city block was cleared for urban renewal. Deep excavation involved a three-storey basement over the whole site and four deep shafts of 23m and 25m diameter for large caisson foundations resting on very dense soil. Diaphragm walling and steel bracing were used for the excavation for the basement. The shafts were excavated making use of diaphragm walls in a circular layout. The circular action of the wall obviated the need for struts, but ring beams were constructed at intervals to allow for the bending effects of non-verticality of individual wall panels. Since then, several other circular types of layouts have been adopted for diaphragm walling including larger circles, figure of 8 double circles and the "Caterpillar" with four over-lapping circles.

5.3.1.11 A stitch in time

A deep excavation, say for an eight-car underground railway station, removes about 200,000 tonnes of the ground. The ground is compressible, and when it is unloaded it swells. The amount of swelling depends on the load removed and the swelling properties of the ground. When the structure is built, load goes back onto the ground. In the case of piling, the imposed load is carried down below the formation level, but no matter how small, there is some movement. When the groundwater table is near the surface, basements tend to be buoyant and might even have tension piles to hold them down. On the other hand, the superstructure usually ends up imposing a load and piles under the superstructure are in compression causing the ground to settle. Base-ments that extend beyond the footprint of the superstructure will move differ-ently than the superstructure during construction as these changes in loading take place. If the differential effects are small, the basement and superstructure can be built as one. However, if the differential movements are large, a gap can be left between the outer basement structure and the portion beneath the superstructure. The gap allows the basement and the superstructure to move independently. When the movement stops, the gap can be stitched together. If more movement is expected in the future, a movement joint might be installed or the stitched connection can be articulated.

This technique has often been adopted for large buildings with deep base-ments, especially where the site is underlain by compressible soil to a great depth. A design for a high-rise office building in Jakarta included a reflective pool for which the tolerance on settlement was a few millimetres. The pro-ject included a deep excavation for basements over the whole site surround-ing a tower block. The ground beneath the site comprised deep layers of soil with intervening layers of lahar, which originates from hardened volcanic debris flow. Field and laboratory testing indicated that the ground would be relatively stiff for small changes in loading but less stiff for larger changes of loading. Piling would be necessary for the tower block. It was estimated that bulk excavation for the basement would result in sufficient unloading to reduce the stiffness of the ground by too much. Likewise, it was estimated that imposing the whole weight of the intended building would impose sufficient load to reduce the stiffness of the soil too much. It was estimated that if the building could be the wished-in-place, whereby the net weight of the building minus the excavation is imposed on the ground, the stiffness of the ground would remain at a high value and settlement would be small. the deflections of the ornamental pool would be within the required narrow limits. Wishing into place is an ideal. However, by phasing the excavation and construction, changes in loading and unloading could be reduced and the stiffness of the soil could be kept high and settlements could be reduced. Large diameter piles were driven from the surface before bulk excavation. The piles were used at first to hold down the soil during bulk excavation. The bulk excavation then commenced with a limited excavation for the base of the core of the tower at the centre of the site. The core was built and its weight of offset the weight of soil that had been removed. The excavation was then widened for construction of more structure. Unexcavated substantial berms were left around the perimeter of the site. When the structure of the tower and some of the basement were nearly complete, the remainder of the bulk excavation took place. The structures were completed in stages as the remainder of the basement was excavated. Gaps were kept open for the part structures to settle, or heave, differentially, and when the construction was completed the gaps were stitched together and made watertight. All went to plan at first, and most of the core was constructed, but work stopped during a down-turn in the economy, and the scheme was later changed.

5.3.1.12 Groundwater relief

Uplift of the base of an excavation can occur if the ground water pressure below the base of the excavation exceeds the weight of soil above it. This can happen, for example, when there is a low permeability soil at the final level for the excavation and there is a conductive layer beneath it. One technique to maintain a factor of safety in such a case is to relieve the ground water pressure from beneath the low permeability soil until it is sufficiently less, by the required FoS, than the vertical total overburden pressure of the soil from the permeable layer up to the excavated level. In order to relieve water pressure, the method is to sink wells down to the elevation where the water pressure is to be dropped and extract water from the wells until the water pressure in the ground below is acceptable. A simple way to control the water pressure in the well is to extend the well liner upwards to the required elevation where incoming water can overflow into a drain and by another pump to the surface. There are five things to watch out for. First, if the permeable ground below the base extends a long way it might yield so much water that it is not feasible to lower the ground water pressure by installing many wells. For example, a gravel layer extending to a river would not be dewatered sufficiently. Second, the required drawdown should be verified midway between relief wells and not at the wells because there will be higher pressure between the wells as the ground water flows towards the relief wells. Third, there may be extensive drawdown outside the site which could lead to undesirable settlements. The drawdown of ground water pressure and settlement should be monitored outside the site. Fourth, there must be a back-up plan in case the relief wells do not function properly. Fifth, and probably the most important, test the system before excavating to a depth where the full ground water pressure becomes critical. Testing can be done by installing some of the relief wells at an earlier stage of the excavation and testing the wells. If the water level in the well can be lowered and maintained by a given rate of pumping, the system is feasible and the rate of pumping will indicate how many relief wells are needed. For smaller deep excavations in sensitive locations, a full-scale pumping test is recommended whereby wells are installed and pumped to achieve the required degree of drawdown. Piezometers are located outside the walls to check whether the walls are leaking. If walls are leaking unacceptably, seal the walls from the surface before excavating. Before excavating the ground, water is not flowing, and grouting would be more effective. The main criterion of acceptable leakage is governed by the drawdown that would result and cause settlement outside the site and should be limited. Rate of flow of water must be limited in the presence of dispersive soils that would be mobilised by the flow of ground water into the excavation. Gaps in retaining walls in dispersive soils must be plugged before reaching them and before experiencing inflow of water and soil into the excavation, which is usually fast with severe consequences of rapid settlement and extensive loosening of the ground outside the excavation site.

5.3.1.13 Excavation under water

If relief wells cannot be used, some other method has to be adopted, such as excavating under water with the excavation flooded until the final excavation level is reached and then using tremie concrete to construct a sufficiently thick un-reinforced concrete slab to, with the soil immediately below it, resist the uplift of the ground water pressure at the lower depth. Tremie concrete is placed under water via a tremie pipe that discharges at the bottom of the excavation. The bottom end of the tremie pipe has to be kept under the top of the wet concrete to prevent concrete from mixing with the water and thereby becoming weaker when set. Mass concrete can be placed below water using grabs only if the concrete is required for its weight. An alternative is to use rock rubble filling material placed by grabs across the base of the excavation under water, but rock fill is not as heavy as concrete, and a greater thickness would be required. For deep excavations in weak soils, water alone can be insufficient, and some bracing can be used.

For example, the excavation can be progressed with bracing until a depth is reached where further excavation would lead to unacceptable deformations of the retaining walls or exceed the practical capacity of the retaining walls. From then on, the excavation could be flooded as described above but with bracing in place. In order to prevent accidentally damaging the bracing, the bracing should remain exposed when the lower portion of the excavation is flooded.

5.3.1.14 Jet grout slab

Another method of preventing base heave due to excessive ground water pressure below the base is to construct jet grout slabs, as described above, from the ground surface before bulk excavation. In this application, the jet grout slab not only braces the retaining walls at depth before bulk excavation but also resists uplift of the base. Jet grout at overlapping spacing can form dense plain concrete with strength of 5MPa or even 10MPa. Some designers make use of such a slab to resist heave by arching across the width of the excavation. If arching is relied on, the slab should be checked for shear. If there are gaps between individual JGPs, the shearing resistance can be reduced to the adhesion of jet grout with the remaining soil between the JGPs. The adhesion with the soil can be very low. Jet grout drilling machines are relatively lightweight, and the verticality has to be checked regularly. For systematic jet grouting, the setting out of the drilling locations is important, and the process can be made more accurate and sped up if rails are fixed and the drilling machine is put on a platform with castors to be pushed along the pre-set line from hole to hole.

5.3.1.15 Piping

Piping is another potential problem with ground water pressure beneath the base. If the soils that are exposed at the final excavation level are permeable and dispersive, there is a risk that piping failure could occur. The phenomenon is that a little water seeping upwards can start to erode the dispersive soils; then the water pathway widens until there is an unmanageable inflow of soil and water. This is the proverbial Dutch boy not putting his finger in a small hole in a dyke and saving the village from flooding. If it occurs, the immediate remedy is to block the inflow with bags of soil if the rate of inflow is slow enough or to use earth moving equipment to place enough loose soil over the affected area to stem the inflow. Of course, excavation cannot resume until the situation is remediated. Prevention or remediation requires reducing the upward flow of water. Where feasible, relief wells can be used to lower the water pressure below the base of the excavation. Otherwise, a better hydraulic cut-off around the site should be provided, such as by grouting directly beneath the retaining walls to make the ground there less permeable. Therefore, it is necessary to consider the potential for base heave and piping when assessing the ground water regime and the properties of the soil beneath the final level of excavation.

5.3.1.16 Compartments

When the base of an excavation is weak, the lateral pressures from the embedded portion of the retaining walls might cause the formation to heave. When there is positive ground water pressure beneath the formation level, the water pressure can cause the formation to heave. Excessive lateral pressures can be resisted by additional bracing such as more struts or such as Jet Grout slabs as discussed in Section 2.1.9. However, excessive groundwater pressure is more of a problem. When there is an aquitard, such as a layer of clay at or below the formation level, ground water pressures in the underlying stratum might exceed the deadweight of the aquiclude up to the formation level.

Base heave can be overcome to a limited extent by excavating in compartments. The method is to excavate generally to a safe level where the remaining soil resists heave and to excavate further within a limited area thereby creating a compartment extending down to the final level. Within the compartment, the base structure can be built and connected to walls and columns that can provide resistance to up lift, and then another compartment can be excavated. By this means the final formation level and the base structure can be constructed in stages.

5.3.2 Good projects and good planning

The last few decades have seen massive migrations to cities in South East Asia. Other cities have grown in population too. In many cities much of the infrastructure is overloaded, and open space is at a premium. Underground space is mostly an un-tapped resource; but there is increasing use of underground space by metro railways and as downwards extensions of building into basements where many activities can be carried out. Car parking in basements is quite common, but there are also supermarkets and a host of other shops, leisure facilities such as ice-skating rinks, cinemas and tuition centres. In several cities, connecting basements to underground stations has resulted in increasing property values. Many railway stations in Japan have underground precincts and subways allowing pedestrians to roam freely without having to observe traffic lights, with casual meeting areas and connections to basements of department stores and recreation complexes. In Kaohsiung the construction of the metro included a piazza under the roundabout right at the centre of the city where a brilliant art deco glass-domed ceiling with back lighting by Narcissus Quagliata [49] is not just a tourist attraction but also a place for concerts and even for weddings.

In many cases, the chosen method of construction has been by deep excavation and either top-down or bottom-up construction of the permanent structure. Construction underground is often criticised as being expensive, but compared to the value of the sites aboveground in the heart of cities and the lack of space for facilities that must be at the surface, underground space becomes a viable option. Costs for underground construction increase dramatically with complexity, and the first underground construction in a given area could render subsequent underground developments prohibitively expensive. Therefore, planning the development of space below ground is an important activity not only to rationalise and coordinate the uses but to reduce overall costs compared to ad hoc development.

The Singapore Government is probably a leader in strategic planning including the integration of developments above and below ground. Master Plans are for implementation and Strategic Plans; looking 40 to 50 years ahead consider territory-wide options and implications and the direction for the Master Plans to take. In order to facilitate development underground, recent legislation reverted ownership of bedrock below leased and freehold land, in principle from 30m below top of rock, to Government. This is a great step forward since usage of underground space can be planned without too many land ownership issues. Another step is to establish a geological database holding information about the ground that can be used for planning and a digital record of underground utilities that can be accessed and interrogated using a Building Information Management (BIM) system.

For the new reclamation area at Marina Bay South, rules for building development require basements to be built with provision to link to adjacent basements, via pedestrian subways to underground railway stations and underneath streets to other basements. Some of these pedestrian subways and meeting areas have already been built. It is planned that new streets will be equipped with combined utility tunnels with access for maintenance and repair with the expectation that the streets will not be excavated by the utility owners nominally for one hundred years. For major projects to go ahead, stages of the review process include investigating the potential for underground space. How much of the facility could go underground? How much more space underground could be created, and could there be co-development or access for further development in the future? In Hong Kong, preliminary design is under way for a new waste water treatment plant for about one million people to be built underground, actually in caverns in rock, so that the existing plant can continue to operate until the new plant is tested and the flow is switched over. The site of the existing plant can then be cleaned and put to a more beneficial use such as housing.

In Singapore, Jurong was developed as an industrial area, "out of sight and out of mind." Over the last decade or so, it has been transformed into a New Town. New commercial buildings have as many as six floors of basements and, unless one looks at the floor numbering one cannot tell whether one is above or below ground. The basements of adjoining lots have been interconnected whereas the main connector above ground is an extensive system of footbridges called the "J-walk," Once indoors, one can spend a whole day in air-conditioned comfort.

The essence of Singapore's planning is to make full use of all of the land surface and the developable space above and below ground, and as a consequence Singapore has constructed many deep excavations. Good planning leads to good projects.

5.3.3 Dealing with geological risk

Unlike manufactured materials produced to specifications, tested and quality assured, the ground is a natural environment and at the very start of a project an unknown factor. The geological risk for deep excavations and tunnels can be high. The probability of encountering unforeseen ground conditions can be high and the consequential costs can be very high. Claims for unforeseen ground conditions can equal the value of the contract for deep basements and can be double the value of the contract and more for tunnels. Therefore, it is important that risk should be evaluated at all stages of a project and mitigation measures considered, evaluated and adopted where appropriate. At the start of a project the site has to be investigated, the types of ground beneath the site have to be identified and the engineering properties for each type of ground that will be affected by the proposed works have to be quantified. Site-specific GI usually is based on vertical boreholes with in-situ testing and sampling. These can be supplemented by profiling between borehole locations by sensing such as geophysical, resistivity and micro-gravity surveys. The more information that is obtained, the lower the uncertainty about the ground conditions. The results from GI are mostly assessed subjectively. Sometimes in well-developed areas sufficient data is available to evaluate parameters about the ground statistically. Standards of design are set with FoSs in order to mitigate risk. Measures such as supervision, instrumentation and monitoring the performance of the works and inspection of the standard of completion are adopted and limits are set for protection of the works and the surrounding area and property.

There are means of managing geological risk. In some cases, it is possible to quantify geological risk for new projects; sometimes geological risk can only be evaluated qualitatively. Quantified Risk Assessment (QRA) relies on performance data. For example, the oil industry has practised QRA for a long time and gathered performance data such as the frequency of failure of a given type of valve so that the probability of failure of many components in a system or in the whole production plant can be summed. Approving authorities set low probabilities of occurrence that shall not be exceeded, and insurers evaluate the financial risk and thereby determine the value of premiums to be paid.

5.3.3.1 Planning

Managing geological risk begins at the planning stage with gathering as much data as possible about the site including published geological information and historical records about the site. Available information includes geological maps. At some places, only regional maps are available, which indicate the general geological conditions such as type of rock and approximate traces of major faults. At other locations, the geological maps that are available can be very detailed. For example, in Hong Kong the Hong Kong Geological Survey has gathered geological information from many projects and has produced geological maps to a scale of 1:5,000 and geological memoires to even smaller scales. In locations of intensive development, archives of borehole records and ground investigation reports can provide a lot of useful information about the ground conditions in the vicinity of the site. For re-development, there may even be historical boreholes from the previous development. In Hong Kong, the GEO established a GIU that has collected geological and geotechnical information useful for planning deep excavations and other ground engineering works. Their database of archived boreholes now exceeds 400,000. In addition, as-built record plans are archived by BD. Plans include as-built foundation records, many of which give information about the elevation of rock at the location of the building. A word of caution, some of the old records may not be accurate, as explained in Section 3.6.2. Caveat emptor, buyer beware, there is no guarantee attached to borings dating from those days.

5.3.3.2 Ground investigation

GI including site-specific boreholes is usually carried out for the project. Typically, a sample diameter is about 76mm. Spacing of boreholes at the design stage might be 10m for a basement in a city block or more widely spaced such as to 50m for an underground railway station depending on the anticipated ground conditions. The plan area of one borehole is about 0.0044 m^2. At 10m spacing, one borehole is about 0.0044% of the ground. If samples of soil 0.3m long are taken every metre, the frequency of sampling by volume is 1 in 75,000. This is quite a small degree of sampling. The reliability of the sampling depends, of course, on the degree of uniformity of the ground. For large geological features such as the sediments of the Taipei Basin, it is well known that the alternate lacustrine deposits of clay and sand are extensive across the basin. The extent of the presence and thickness of the respective layers depends on the relative historical dominance of the Tamsui and Keelung rivers, within distances such as one kilometre the types of soil and the thickness of the alternating sand and clay layers does not change by very much. It is arguable that a thorough ground investigation of an adjacent site could be used for the project at hand.

For locations where ground conditions change rapidly, there is a higher geological risk than in the relative tranquillity of lake deposits; more detailed site-specific ground investigation might be required. For example, in topically weathered rocks the rock head can be extremely variable as discussed in Section 3.6.3. Since excavating in rock instead of soil can take ten times as long and cost perhaps ten times as much, the percentage of rock to be excavated with diaphragm walling equipment is a geological risk. Likewise, the amount of rock within the bulk excavation for a deep excavation has serious consequences in cost and time and possibly the need to mobilise more or different equipment if the amount of rock to be excavated is more than expected. There have been many claims taken to arbitration concerning unforeseen amounts of rock in excavations for diaphragm walls and bulk excavation.

5.3.3.3 Basis of design

The next step in mitigating geological risk is the margin of safety that is included as a basis of design. Codes of practice and regulations specify FoSs, or partial factors on loads and on material to ensure that failures "do not occur." In reality, the objective is to achieve serviceability under normal conditions for 100 years or thereabouts and an extremely low probability of failure under extreme events. Lower FoSs can sometimes be adopted during construction because the site is not open to the public; it is occupied by people who are, or should be, briefed as to the activities and safety concerns; the works are, or should be, monitored under technical surveillance; and there is an AAA system in place to alert responsible and technically competent people if there is a trend that might lead to untoward movements.

As discussed in Section 5.2.3.2, the COI, following the collapse of Nicoll Highway in Singapore, recommended adopting FoSs for temporary deep excavation that had no reduction from the FoSs required for permanent works. This was a response to a recognition that deep excavations have a high likelihood of collapse. Risk involves hazard and consequence. Hazard relates to the threat or probability of occurrence, and consequence is the impact of the event, such as physical damage or fatalities. This is evident in the *"Manual for Slopes"* issued by GEO [50], where the risk for slopes are graded according to the geotechnical hazard and the consequence of failure, for example low consequence within a country park or high risk alongside a school.

Codes of practice and regulations do not go as far as to quantify the method of selection of values for parameters for soil. Global FoSs envisaged typical values that are expected to pertain nearly always. Limit State codes such as Eurocode 7 [15] propose moderately conservative values for geotech-nical parameters to be used in conjunction with material factors. What is moderately conservative is not defined. It remains for the designer to evaluate the values for him- or herself. If enough data is available, then statistical methods could be adopted. Gradually more and more data is being published. However, using this method to estimate the top 1%, ten results out of 1,000, might be a means of determining the 98th percentile above the mean value, and cutting off the bottom ten results would give the statistical 98th percentile below the mean value, but one needs to be cautious about the reasons for weakness. In zones of decomposition of rock there can be up to two orders of magnitude range in strength for fresh rock to highly weathered rock from, say, 300MPa to 3MPa strength and the percentile should be calculated for the specific grade of rock and not necessarily for all grades of decomposition.

5.3.3.4 Conditions of contract

The next stage where risk mitigation can be adopted is in drafting contract documents. In these documents, the allocation of geological risk between an employer and a contractor should be identified. Two decades ago, many con-tracts were drawn up with a clause whereby a contractor should take all the geological risk. An employer expected to have fixed prices or firm rates for getting the work done and the project completed. Civil engineering contracts were frequently tendered based on competitive pricing. This was not a satisfactory arrangement because a tenderer needed to get the lowest price, which meant not only competitive procurement but also a low budget for geological risk. Projects such as tunnelling and deep excavations have large geological risks and the prospect of large cost over-runs. Many substantial claims for unforeseen ground conditions resulted, and costs of arbitration or court hearings considerably added to the costs.

A simple means of mitigating risk is to endeavour to set a reasonable period for the duration of the contract. If the contract duration is very short, then the consequence of unforeseen events can be difficult or even physically impossible to overcome within the short contract period.

GBRs were instituted to define what range of values for baselined parameters would be foreseen by the contract. The objective was to pro-vide a basis for tendering in order to achieve competitive tendered prices and definitions of the range for geotechnical parameters that the success-ful contractor should expect to encounter and to provide the baselines that an engineer would take into account when evaluating claims for unforeseen ground conditions. The baselines do not foresee the quantities of materials (e.g. how much very strong rock and how much very weak rock is foreseen by the contract) only that strength up to the maximum or down to the minimum are foreseen. Foreseen in this context means envis-aged in drafting and administering the contract. Foreseen does not imply that such conditions, high or low, will actually be encountered. GBRs have been used for several contracts in Hong Kong and in Singapore. When properly drafted, GBRs have been proven useful. In arbitration, some GBRs appear to have not been well drafted. When enough data is available, baselines can be set in accordance with, say, 95% probability of occurrence or even the whole range of the available data plus a margin. In this way, an employer can gauge the financial risk. In most cases, the GBR allows for evaluation of claims for cost; conditions of the contract do not usually allow for variation of time for completion. With respect to unforeseen ground conditions, a contractor accepts the time-based risk with a need to accelerate or change his method of working to cope, and an employer accepts financial risk for reimbursement of actual costs. There is a sharing of risk: risk on time is assumed by a contractor; risk on cost is assumed by an employer. An alternative method of reducing a contractor's geological risk is to re-measure, which means that the contractor gets paid according to contracted rates or sums for every item of work that is certified as having been done. There should be a rate for every type of work, but when unforeseen ground is encountered there might be disputes about the rates. Re-measurement for excavation, whether in diaphragm wall trenches or in bulk for any type of material, addresses changed depths for diaphragm wall panels because the dimensions of the excavation are usually finalised. Re-measurement for excavation in ground and extra over for rock is normally sufficient for a fair compensation for work done.

Civil engineering contracts generally address the geological risk, but most contracts are awarded based on lowest price in a competitive tender. This puts pressure on construction companies to reduce the allowance for risk, and many arbitrations have resulted when contractors have sought recovery of costs that they had not allowed for when pricing the job.

Other forms of contract, such as Target Cost and Partnering are aimed at fair resolution when encountering unexpected circumstances [51]. Other forms of contract, for example the architectural forms of contract are not so good. Architectural work involves mainly manufactured materials and products produced to a standard, and variations in costs relate more to fluctuations in costs of basic materials, fuel and labour. Very often geological risk affects only part of the project, for example the foundations and the basement, and is not addressed or is assumed totally by a contractor.

5.3.3.4 Assessment for tendering

The next stage is when the contract for construction is tendered. The tenderers need to understand the ground conditions in order to prepare estimates for cost and time and any risk. It is important that tenderers have access to as much information about the site as possible. For example, borehole data and geotechnical reports are site information. Without such data, the tenderers are much in the dark and can only see the current condition of the site at ground level. In the past, some employers refused to hand over factual data in case it was considered incorrect or not representative and if a contractor had relied on the data, he would claim the consequential costs. The law courts in the U.K. take a dim view of employers who withhold factual data. For large projects, it is now a practice to prepare a geological model using a numerical database. Handing such a model over to the tenderers saves them a lot of time when preparing their tender. It is far quicker to check such a model than it is to compile one. It is a far better practice than allowing inspection of hard copy reports and limiting the taking of copies to a small number. Very often, the tender period is quite short, and it is difficult for a tenderer to assess the data that he can obtain within such a short period of time. Handing over a GIR would facilitate the tenderers but interpretative reports that are prepared for purposes of design are not necessarily good interpretations for planning construction. The differences of interpretation might be subtle but in addition to any interpretation for purposes of design, the contractor should have an interpretation for purposes of construction. Many contracts require the contractor to submit his own interpretative report at an early stage.

5.3.3.5 Handing over information

For many projects, there is a lot of site data gathered by the designer which the tenderers need to assess before submitting their bid. Such data includes a model of geological and geotechnical data about the ground beneath the site as well as utilities and nearby structures. A convenient way of transferring this data to tenderers and for the tenderers to interpret is in a BIM. During construction, further information such as monitoring instrumentation can be added and potential clashes, say for tie-back anchors and utilities, can be identified and avoided.

5.3.3.6 Sharing risk

As mentioned above, in 1990s deep sewer tunnel contracts in Hong Kong went to arbitration for claims of unforeseen ground conditions. There was a large settlement out of court in favour of HKSAR Government, HK$750 million was reported [52] and there were replacement contracts to complete the tunnels. An international panel of experts made recommendations aimed at mitigating the risk of such claims. These included more ground investigation and equitable sharing of geological risk. In preparing contracts for further deep sewer tunnels, more ground investigation was carried out; a geological model was prepared for the project and made available to the tenderers. Key cost items for probing the ground ahead of tunnelling and for grouting the ground to limit inflow of groundwater were re-measured. At the same time, drainage tunnels were constructed for the same employer and GBRs were used. For one of the drainage tunnel contracts strong granite rock, mixed ground of soil and rock and soil were baselined (published data is not yet available).

5.3.3.7 Probing

In order to reduce the uncertainty about the conditions of the ground, further GI can be undertaken soon after the start of the contract. GI after the start of a contract is usually specifically geared to the uncertain parameter. Probing is often used. For example, as discussed in Section 3.6.9, when rock is expected in excavating trenches for diaphragm wall panels, the occurrence of rock is a risk item. In Hong Kong it is common practice to probe by pre-drilling each panel. Usually the hole is drilled, and the rate of progress is quite fast in soil. Rock is recognised by the reduced rate of drilling. When rock is massive, probing is often stopped on first encountering the rock. In weathered rocks where corestones and weathered seams below rock head could be present, probing is sometimes continued to record the presence and absence of rock material by the rate of advance of drilling, or drilling could revert to coring. Both methods would be aimed at measuring the percentage of rock encountered.

5.3.3.8 Inspection of the ground

During construction the assumptions adopted for design are verified by inspection. If the ground conditions are found to be different, the design is reviewed, and any necessary changes are made.

5.3.3.9 Monitoring the performance

Monitoring is adopted to verify that the performance of the ground and the structures are as predicted by the designer. AAA limits are set to control the work. If the results of monitoring are different from the predictions, the design and method of construction are reviewed; if necessary, they are changed.

5.3.3.10 Inspection and supervision

It is often good to solve problems quickly. When changed circumstances are discovered on site a well-informed and quick resolution between a contractor and the RSS can prevent the problems from escalating.

Escalation entails investment of time and fees for professional and legal services.

5.4 What next

When I first worked on deep excavations in urban areas in the mid 1970s, there were few under construction. Basements were generally of two floors or less. Pioneering for South East Asia, in Hong Kong the MTRC was planning to build a modern underground railway with smooth welded rails and air-conditioning. Computers were of limited capacity, and commercial software was scarce. Since then many cities have built first class underground railway systems, and basements are often six floors deep and sometimes more. As big cities become bigger, more cities develop underground space due to congestion at the surface, and deep excavations are used for a whole variety of functions, such as supermarkets, restaurants, cinemas, libraries, sports and recreation facilities; in places that are hot, such as Singapore and Hong Kong in the summer, people spend weekends in air-conditioning above and below ground.

5.4.1 More deep excavations

There are increasing numbers of deep excavations under way or planned. Urban planners, such as Urban Redevelopment Authority in Singapore and the Planning Department of HKSAR Government include development of underground space in their strategic planning. Many cities are developing underground railways or expanding existing underground railway systems. Property developers have realised the commercial value of deep basements. Connections of basements to underground railway stations provide low cost entrances for the railway system and generate large flows of pedestrians going to and from the trains that are welcome because the basements then become prime sites for retail shops.

Drivers for underground development include the following:

- Congestion in urban areas inflates the price of floor space and renders sites for construction a scarce commodity.
- Increased population creates demand for increased facilities
- Improved incomes cause people to want better facilities for a better way of life.
- There is an awareness that underground space is developable and in many cases is a preferred method.
- Environment can be improved by locating facilities below ground and freeing surface land for open space and recreational use.
- Increased economies make funds available for improvement of city life and functions.

These factors apply to many cities; as a result, there are increasing numbers of deep excavations. Moreover, new excavations are generally deeper. Basements with six floors, about 20m deep, are quite common for new buildings.. In Victorian England, a smart town house would have a single storey coal cellar. With many underground utilities and underground roads and rail, new utilities and transportation tunnels are getting deeper to pass beneath existing underground structures.

5.4.2 Expanding technology

Technology has been expanding rapidly. Mobile computers marketed as android telephones are very versatile. For deep excavations, one can connect via the internet to the site data management system and read data for any period of time for any instrument or set of instruments. In offices, for design and on site, computing systems are very powerful, and what was impossible for a team to calculate in their lifetime can be computed in fractions of a second.

There are more types of instruments for monitoring on site, and intelligent instruments talk to each other. If an instrument is knocked off-line or is faulty, the surrounding instruments sort it out with coordinates for its new location or an alert for replacement. A lot of monitoring is automated now using an IT cloud to make information available, thereby dispensing with old-fashioned sets of thick lever-arch files of paper copies of results every month.

5.4.3 Learning from previous experience

Artificial Intelligence was invented by mankind mimicking human learning from experience. A wise person learns from his mistakes and can learn from mistakes made by others. Section 5.2 addresses lessons to be learned from failures. It is important to the engineering profession that lessons learned are retained and put into practice. Increasingly with powerful aids, such as computers, engineers become engaged with larger projects and with more projects, and lessons run the risk of being forgotten or not passed on to those who could benefit. Looking into the future, the trend is increased responsibility for projects with increased complexity that require increased vigilance and not complacency.

In a broad sense, a lesson learned is experience, and experience includes gathered, published and archived data. The more data that is available, the more knowledge and insight, provided that the collective data is validated and taken into account. A risk for the profession is the number of publications that can be found too easily by using a search engine on the internet. Advice from senior colleagues such as can be obtained by peer reviews is important. Peer reviews should be adopted frequently.

Large libraries of data and information are increasing in accessibility. For example, the GIU in Hong Kong was established 25 years ago and is mostly available on-line. Urban Redevelopment Authority in Singapore is developing a geological database. As more data becomes available, statistical evaluation becomes meaningful and QRA becomes feasible. With so much data available and with many uncertainties about ground conditions, statistical evaluation of a large relevant data set offers a potentially more reliable basis of design than, perhaps, a site-specific investigation restrained by a tiny budget.

5.4.4 New machines, new materials

Excavating the ground has been going on for by hand for thousands of years. After the Industrial Revolution, excavating machines were adopted. The basic method of digging soil and loading into lorries or onto conveyor belts is well practiced. Whereas tunnel-boring machines have been much improved during the last four decades, basic earth movers are much the same. Forty years ago, most of the diaphragm wall panels were excavated by grabs. The invention of Hydrofraise cutters considerably increased rates of excavation in soil and in rippable rock.

Slurry trenches during excavation for diaphragm walls commenced using natural bentonite. Synthetic bentonite and polymers are commonly substituted instead of natural bentonite. However, problems with slurries continue because of the conflicting requirements to suspend excavated soil but to remain clean for support to the sides of excavated trench and during concreting.

Deep excavations extending below the ground water table often require grouts for reducing flow of water and in weak ground grouts are required to strengthen the ground. Several types of grouts are available. However, to some extent grouting is an art and not a technology. Some grouting specialists are worth their weight in gold, but at times mis-information has led to mis-understanding about what can be achieved such as by pumping cement slurry into the ground as discussed in Section 5.2.9.

Concrete with steel reinforcement has been in use for at least 100 years. The two materials have some good qualities: concrete is quite strong in compression, and steel is strong in tension. The two materials also have poor qualities: concrete cracks and steel corrodes. Some reinforced concrete structures have not survived an intended design life of 100 years and have required repair or replacement. Design and maintenance of reinforced concrete structures requires control of cracking and durability against corrosion. Chemical companies produce many additives to make concrete more durable. There are several additives for concrete to improve water-proofing, hardening, workability and so on. Steel fibre reinforcement has benefits for crack control and potentially for durability. Glass fibre reinforcement bars are easier to cut when planned for a TBM to cut through a reinforced concrete diaphragm wall at the end of a deep excavation. Fibre reinforcement bars are often used for *"weak eyes."* The global annual consumption of concrete and reinforcement is vast. and improvements in concrete and reinforcement could have a large impact.

5.4.5 Complexity

Cities grow, becoming large and intensively developed. Increased complexity for deep excavations in the future seems to be inevitable. In developed cities the space below ground is becoming increasingly developed. The best sites have been used already. Urban reconstruction whereby large blocks of old developed land are resumed happens, and the comprehensive development of the blocks enables underground development to be adopted. Urban planning facilitates underground development, and as buildings get older, deteriorate, need replacement and leases expire, it is to be expected that more urban reconstruction will lead to more deep basements in urban areas.

Sites are complicated by adjacent underground structures and adjacent buildings. Increased complexity is to be expected. Historical artefacts can disrupt the work, and existing buried utilities may clash or be affected.

Increased complexity requires a wider range of skills and versatility. Whereas early in my experience some engineers acquired the two disciplines of geotechnical engineering and structural engineering and became ground engineers skilled in both disciplines. Now it is commonplace for a design contract to include several stations and interlinking tunnels as one design package for design and for construction. Increased size of projects and demands for streamlining the processes have tended to result in a need for teams of engineers who are individually specialised in one discipline to be integrated one with another. Consequently, there are engineers who have a limited range of experience, for example reinforced concrete design, with little education or experience of the other skills, for example evaluation of claims or resolution of disputes. Repeatedly working exclusively within their area of increasing expertise such as structural design, they tend to lose insight of the other skills. Continuing professional development is a requirement of most institutions of engineers and, as an engineer gains more experience within one area of expertise, it is all the more important to broaden his or her knowledge of related skills and to refresh memories of lessons learned.

References

[1] LiDAR. www.neptectechnologies.com/autonomous-lidar/?gclid=EAIaI QobChM Ixp3Xv4iO5QIVSr3ACh1c9A60EAAYASAAFgIiz_D_BwE
[2] Geotechnical Information Unit, Geotechnical Engineering Office, Civil Engineering Department, HKSAR Government. www.cedd.gov.hk/eng/public-ser vices-forms/geotechnical/library/index.html

[3] *Guide to Site Investigation (Geoguide 2)* (Continuously Updated E-Version released 18 December 2017). Geotechnical Engineering Office, Civil Engineering Department, HKSAR Government. www.cedd.gov.hk/filemanager/eng/content_108/eg2_20171218.pdf

[4] Knill, J. Sir. Former Professor and Dean, Imperial College, London. www.geolsoc.org.uk/About/History/Obituaries-2001-onwards/Obituaries-2002/Sir-John-Knill-1934-2002

[5] BS 5930:2015. *Code of Practice for Ground Investigations*. BSI, ISBN 978 0 580 800627.

[6] Park Seismic. www.parkseismic.com/Whatisseismicsurvey.html

[7] Keele. (2019) www.keele.ac.uk/geophysics/microgravity/

[8] Sumo. (2019) www.geophysical.biz/res1.htm

[9] Louis Menard, French Engineer. www.menard-group.com/en/history/

[10] BS 5930:2015. *Code of Practice for Ground Investigations*. BSI, ISBN 978 0 580 800627.

[11] ASTM D3441-16. (2016) *Standard Test Method for Mechanical Cone Penetration Testing of Soils*. ASTM International, West Conshohocken, PA. www.astm.org

[12] Robertson, P.K. (2009) Interpretation of Cone Penetration Test- A Unified Approach. *Canadian Geotechnical Journal*, Volume 46 Issue 11, pp. 1337–1355, doi: 10.1139/T09-065.

[13] BS 1377, Parts 1–8, BSI, UK https://shop.bsigroup.com/

[14] Ban on Hand Dug Caissons, APP-59, BA. HKSAR Government. www.bd.gov.hk/doc/en/resources/codes-and-references/practice-notes-and-circular-letters/pnap/APP/APP059.pdf

[15] Eurocode 7. (2013) *Geotechnical Design*, European Union. ISBN 978-92-79-33759-8.

[16] Buildings Ordinance Cap.123. HKSAR Government. www.elegislation.gov.hk/hk/cap123

[17] Utterberry, Trademark. https://utterberry.com/

[18] von Terzaghi, K. (1883–1963) The Father of Soil Mechanics. https://en.wikipedia.org/wiki/Karl_von_Terzaghi

[19] Peck, R.B. (1912–2008) Professor, University of Illinois. www.tunneltalk.com/Obituary-Ralph-Peck.php

[20] Pozzolana. https://en.wikipedia.org/wiki/Pozzolana

[21] Aspdin, J. https://en.wikipedia.org/wiki/Joseph_Aspdin

[22] PNAP-22 Practice Note 74 (2009) *Dewatering in Foundation and Basement Excavation Works*, Building Authority. HKSAR Government. www.bd.gov.hk/doc/en/resources/codes-and-references/practice-notes-and-circular-letters/pnap/APP/APP022.pdf

[23] *Geotechnical Manual for Slopes*. Geotechnical Engineering Office, Civil Engineering Development Department, HKSAR Government. www.cedd.gov.hk/filemanager/eng/content_91/egms.pdf

[24] Telford, T. (1757–1834) Civil Engineer. https://en.wikipedia.org/wiki/Thomas_Telford

[25] Brunel, I.K. (1806–1859) Civil Engineer. https://en.wikipedia.org/wiki/Isambard_Kingdom_Brunel

[26] Tay Bridge Collapse. https://en.wikipedia.org/wiki/Tay_Bridge_disaster

[27] Tacoma Narrows Bridge Collapse. https://en.wikipedia.org/wiki/Tacoma_Narrows_Bridge_1940

[28] Aerodynamic Flutter. www.ketchum.org/billah/Billah-Scanlan.pdf

[29] Martell Bridge: "UK Military Bridging – Equipment (Pre WWII Equipment Bridging)." www.thinkdefence.co.uk. Retrieved 8 May 2017.

[30] Committee of Inquiry into the Basis of Design and Method of erection of Steel Box-Girder Bridges. (1973a) *Inquiry into the Basis of Design and Method of Erection of Steel Box Girder Bridges; report of the Committee*, H.M.Stationery Office, London.

[31] Report of Royal Commission into the Failure of the West Gate Bridge. (1971) Government Printer, Melbourne, www.parliament.vic.gov.au/papers/govpub/VPARL1971-72No2.pdf

[32] Committee of Inquiry into the Basis of Design and Method of erection of Steel Box-Girder Bridges. (1973b) *Inquiry into the Basis of Design and Method of Erection of Steel Box Girder Bridges. Appendix 1: Interim Design and Workmanship Rules*. H.M. Stationery Office, London.

[33] BS5400, Steel, Concrete and Composite Bridges. BSI, UK ISBN 0 580 10185 1.

[34] COI Report. (2005) *Report of the Committee of Inquiry into the Incident at the MRT Circle Line Worksite that Led to the Collapse of the Nicoll Highway on 20 April 2004*. Ministry of Manpower, Singapore. (Call no.: RSING 363.119624171 SIN).

[35] Roth, W. (2018) Vice President and Principal Engineer, AECOM, Los Angeles, CA. Private communication.

[36] Advisory note 1/05 on Deep Excavation. (2005, May 5) Commissioner for Building Control, Government of Singapore.

[37] Endicott, L.J. (2013) Examination of Where Things Went Wrong, Nicholl Highway Collapse, Singapore. *7th International Conference on Case Histories in Geotechnical Engineering*. Wheeling, IL April 29–May 4; Endicott.

[38] Endicott, L.J. (2013) Lessons Learned from the Collapse of the Nicoll Highway in Singapore April 2004. *IABSE Conference*, Kolkata, September.

[39] Tattersall, J.W., Tam, J.K.W., Garshol, K.F., & Lau, K.C.K. (2012) Engineering Geological Approach for Assessment of Quantities and Programme of Deep Tunnels in Hong Kong. *Annual Seminar, Geotechnical Division Annual Seminar*, Hong Kong Institution of Engineers.

[40] Legislative Council Secretariat: Information Note IN26/02-03: The Legislative Council Building.

[41] *Guide to Site Investigation (Geoguide 2)*. (Continuously updated E-version released 18 December 2017), Geotechnical Engineering Office, Civil Engineering Department, HKSAR Government.

[42] Endicott, L.J. & Zhang, W.G. (2019) Numerical Evaluation of Top of Rock Profile, Rock Strength and Conductivity. *Proceedings of 44th Conference*, Deep Foundations Institute, Chicago, IL, October.

[43] Pickles, A. (2019) Director, Golder Associates, Hong Kong.

[44] Shakespeare, W. Julius Caesar Act III Scene II.

[45] Endicott, L.J. (1980) Aspects of Design of Underground Railway Statio.ns to Suit Local Soil Conditions in Hong Kong. *Hong Kong Engineer*, Volume 8, pp. 29–38.

[46] Bezzano, M., Smith, S., Yiu, J., & Wiltshire, W.M. Case Study: Design and Construction Challenges for Admiralty Station Expansion. *Proceedings of the HKIE Geotechnical Division Annual Seminar 2017*. www.hkieged.org/down load/as/as2017.pdf

[47] Cheuk, J.C.Y., Lai, A.W.L., Cheung, C.K.W., Man, V.K.M., & So, A.K.O. (2013) The Use of Jet Grouting to Enhance Stability of Bermed Excavation. *18th International Conference on Soil Mechanics and Geotechnical Engineering*, Paris, pp. 1255–1258. www.civilax.com/soil-mechanics-geotechnical-engineering/

[48] Sakaeda, H. & Ozgur, O. (2017) Recent Developments and Risk Management in Immersed Tunneling Projects. *Proceedings of 16th Australasian Tunnelling Conference*. ISBN 9781922107985.

[49] Narcisus Quagliata: Artist Is Stained Glass. www.narcissusquagliata.com/

[50] *Geotechnical Manual for Slopes*, 2nd Edition. (1984) Geotechnical Engineering Office, Civil Engineering Development Department, HKSAR Government, Hong Kong. www.cedd.gov.hk/eng/publications/geo/geo-gms/index.html.

[51] Moorwood, R., Scott, D., & Pitcher, I. (2008) *Alliancing, a Participant's Guide*. Maunsell/AECOM, Brisbane, Australia. ISBN97-0-646-50284-7.

[52] Lawmakers Slam $750m Payout. (2001, October 5) South China Morning Post, Hong Kong, p. 1.

Afterword

This book has been compiled to address all the activities and skills needed to successfully complete deep excavations in soil. It describes the procedures from planning beforehand to resolution of disputes afterward. The component activities are introduced, and references are provided for further insight. Engineers seeking advancement to leadership should be aware of everything in the book. Practicing engineers should take note of the many activities and skills that are required and understand how their role fits in with others in their team whether it is in the design office or on site. It is not necessary, and maybe not possible, to know everything about everything. This book is not intended to provide everything that there is to know; it is aimed at introducing what there is to know and to assist engineers in finding out what they may need to know.

Index

Printed in the United States
by Baker & Taylor Publisher Services